THE ETHICAL ALGORITHM

Safeguarding Humanity in AI

By
Alex Morgan

i

THE ETHICAL ALGORITHM

Safeguarding Humanity in AI

CONTENTS

INTRODUCTION

Welcome to a journey into one of the most significant discussions of our time—ethical considerations in artificial intelligence. As AI technologies continue to evolve at an unprecedented rate, their implications for society have become a focal point of both excitement and concern. This book aims to shine a light on the ethical, privacy, and fairness challenges that AI presents, offering not just a roadmap for understanding these issues but also actionable strategies for navigating them responsibly.

The significance of AI's ethical landscape can't be overstated. The decisions being made today in the realms of AI development and deployment will shape our social norms, economic systems, and even the fabric of our everyday lives. Policymakers, developers, researchers, and tech enthusiasts alike need to grasp the multifaceted nature of the ethical dilemmas posed by AI. This comprehensive guide has been crafted to serve as your indispensable resource in this endeavor.

Throughout history, technological advancements have often come with a mix of improvement and disruption. The printing press democratized knowledge but also displaced the scribes who meticulously copied texts by hand. The Industrial Revolution accelerated productivity while simultaneously igniting debates on labor rights and economic equity. Similarly, AI is not just another technological leap; it's a paradigm shift that compels us to rethink our ethical boundaries, legal frameworks, and social contracts.

In examining the ethical challenges associated with AI, it's essential to start with foundational questions. What is AI, at its core? How has it evolved, and what early ethical considerations have shaped its trajectory? This book embarks on this exploration from a historical and technological standpoint, establishing the groundwork necessary for delving into more complex issues such as privacy risks, bias, and accountability.

Privacy and data protection form a critical axis in the ethical discourse surrounding AI. As data has become the fuel that powers intelligent systems, understanding the balance between utility and intrusion is vital. The types of data collected, the methods of collection, and the ways to mitigate privacy risks are all components that will be rigorously explored. This book navigates these murky waters to offer a clear-eyed view on safeguarding individual privacy while harnessing the potential of data-driven innovations.

No discussion about AI ethics would be complete without addressing the biases that can be unwittingly coded into algorithms. Bias, in many ways, is a reflection of the imperfections found within human society—our prejudices, stereotypes, and systemic inequities. In computational terms, bias can manifest in various forms, from training data disparities to algorithmic structures. Detecting, understanding, and mitigating these biases is a Herculean task but one that is indispensable for ensuring fairness and equity in AI systems.

Transparency and explainability in AI are more than just buzzwords. They embody the principles of accountability and trust. As AI systems govern more aspects of our lives—from loan approvals to medical diagnoses—the demand for clarity about how these decisions are made has never been higher. This book outlines the necessity for clear, explainable models and provides insights into how we can communicate AI decisions effectively to the public.

Accountability is another cornerstone of ethical AI development. Defining who is responsible when an AI system goes awry involves both legal and ethical standards, as well as case studies from different industries. These real-world applications offer concrete examples of how accountability can be implemented and maintained, providing a roadmap for future innovations.

The impact of AI on the workforce is a topic replete with both opportunities and challenges. Automation has the potential to improve efficiency and productivity, but it also raises questions about job displacement and economic inequality. Ethical considerations and preparatory strategies for an AI-driven workforce are key discussion points that this book will thoroughly examine. Ensuring that the benefits of AI are equitably shared requires a collaborative effort among various stakeholders.

In healthcare, the stakes couldn't be higher. AI technologies promise groundbreaking advancements in diagnostics, treatment planning, and patient care. However, these benefits come hand-in-hand with ethical dilemmas and potential risks. This section will showcase case studies and best practices that highlight both the incredible potential and the substantial challenges of integrating AI into the medical field.

The robustness and security of AI systems are often overshadowed by more glamorous discussions but are no less critical. Threats to AI systems, from data breaches to adversarial attacks, pose serious risks. Ensuring the robustness and security of these systems is a non-negotiable aspect of responsible AI development, and this book delves into the best practices for achieving this.

Law enforcement applications of AI introduce a different set of ethical considerations, particularly around surveillance and privacy. The balance between public safety and individual freedoms is delicate and fraught with potential for misuse. As we explore the ethical frameworks and guidelines for responsible use of AI in law enforcement, the goal will be to foster a dialogue aimed at protecting civil liberties while enhancing societal security.

On a global scale, AI's implications transcend borders. Different countries are approaching AI with diverse perspectives and regulatory strategies. This book examines these international viewpoints and offers policy recommendations that aim for global cooperation in AI ethics. The future of AI will be shaped not just by what we do within our borders, but by how we collaborate across them.

Human rights in the age of AI is another realm demanding focused attention. Defining and protecting these rights, especially for vulnerable populations, requires a new framework that can accommodate the complexities introduced by intelligent systems. From humanitarian efforts to everyday applications, AI's role in upholding human dignity and equality is a critical area of exploration in this book.

Finally, designing ethical AI systems encompasses principles and guidelines that can be woven into the fabric of AI development from the ground up. This isn't just about avoiding harm; it's about proactively embedding ethical considerations into the very DNA of AI technologies. Real-world applications and principles of ethical design will offer practical insights for developers committed to responsible innovation.

The ethical challenges posed by AI are manifold, but they are not insurmountable. This book serves as both a map and a guidebook for navigating these complex terrains, offering not just theoretical insights but also practical solutions. As we venture into the nuances of AI ethics, privacy, and fairness, the ultimate aim is to foster a deeper understanding and promote responsible stewardship of this transformative technology.

CHAPTER 1:
FOUNDATIONS OF AI ETHICS

The dawn of artificial intelligence has brought with it a treasure trove of possibilities and challenges, raising an overarching question: How do we ensure the ethical development and deployment of AI? As we delve into the foundations of AI ethics, it's crucial to trace the lineage and ambition behind this technology. From early considerations to its historical context, the narrative of AI isn't just about technological achievements, but the profound ethical concerns it stirs. The lineage of AI urges us to examine how our predecessors have grappled with moral dilemmas and societal impacts. These foundational questions serve as the bedrock for understanding what it means to create and use AI responsibly. Through a nuanced exploration of what AI is and its early ethical considerations, we lay the groundwork for discussions about privacy, bias, transparency, and beyond. Each concept builds upon a shared responsibility and a collective vision to foster AI systems that enhance human well-being while safeguarding fundamental values.

What Is AI?

Artificial Intelligence, or AI, often captures the public imagination through depictions of futuristic robots and autonomous systems. However, AI is fundamentally about replicating human-like intelligence in machines. It encompasses a broad range of technologies that allow computers to perform tasks that traditionally required human cognitive processes. These processes include learning from data, recognizing patterns, making decisions, and responding to complex environments. AI systems can be designed to perceive their surroundings, reason through complex tasks, and even interact with humans in a way that appears conversational.

The essence of AI lies in its underlying technologies and methodologies. From machine learning and neural networks to natural language processing and computer vision, AI draws upon a diverse set of techniques. Machine learning, for instance, involves algorithms that enable computers to learn from and make predictions based on data. This makes it possible for AI systems to improve over time without being explicitly programmed to perform every task.

It's essential to understand that AI is not a single technology but a collection of methods designed to perform specific tasks. While some AI systems are highly specialized, others aim for generalization, exhibiting human-like versatility. Narrow AI, also known as "weak" AI, is adept at performing a single task, such as facial recognition or recommendation filtering. On the other hand, Artificial General Intelligence (AGI) remains a long-term goal and hypothetical construct that would be capable of performing any intellectual task that a human can do.

AI systems are rooted in sophisticated algorithms that carry out calculations and data processing akin to human decision-making processes. These algorithms derive their efficiency from vast amounts of data, often dubbed "big data," and computational power that enables them to process this information at unprecedented speeds. One branch of AI, known as deep learning, uses artificial neural networks to model complex patterns in data, inspired by the human brain's structure.

A pivotal aspect of AI is its ability to mimic cognitive functions. For example, natural language processing enables a system to understand and generate human language, paving the way for applications like chatbots, voice assistants, and automated translation services. Similarly, computer vision allows AI to interpret and make decisions based on visual inputs, which has applications in areas like autonomous vehicles, medical imaging, and security surveillance.

The ongoing development of AI has profound implications for a multitude of sectors, from healthcare and education to finance and transportation. Each of these domains benefits from AI's ability to handle large-scale data, automate repetitive tasks, and offer predictive insights. For instance, in healthcare, AI-driven diagnostics can analyze medical images more accurately and faster than human doctors, while in finance, algorithmic trading systems can execute trades at lightning speed based on real-time market data.

Nevertheless, the growth of AI also raises significant ethical questions. As AI systems become more integrated into daily life, issues such as bias, fairness, transparency, and accountability become increasingly critical. Concerns about data privacy and the potential for intrusive surveillance underscore the need for robust ethical frameworks to govern the development and deployment of AI technologies. Addressing these concerns requires a multidisciplinary approach, combining insights from computer science, law, ethics, and social sciences.

Crucially, the ethical considerations surrounding AI can't be an afterthought. They must be a primary concern embedded in the development process. It's not just about ensuring AI systems perform well; they must also be designed and used responsibly. This means considering the long-term impacts of AI on society and ensuring that these technologies contribute positively to the public good.

From a historical perspective, AI has progressed through several stages, marked by both technological advancements and paradigm shifts. The initial phase, known as the "symbolic AI" era, focused on rule-based systems that explicitly encoded knowledge. This was followed by the "statistical AI" era, which emphasized probabilistic approaches and data-driven methods. Today, we're in the midst of the "deep learning" era, where complex neural networks play a central role in many AI applications.

The future trajectory of AI is both promising and uncertain. While advancements are likely to revolutionize numerous aspects of life and work, they also necessitate careful consideration of the associated ethical dilemmas. As we move forward, the importance of fostering an inclusive dialogue around AI ethics can't be overstated. This dialogue should involve not only experts within the field but also the broader public to ensure that AI development aligns with societal values and needs.

Understanding what AI is and how it functions is the cornerstone for grappling with its ethical implications. It allows developers to create systems that are not only efficient but also ethical. Policymakers, researchers, and the public must work together to create regulations and standards that safeguard against misuse while encouraging innovation. As AI continues to evolve, maintaining an ethical perspective will be vital in harnessing its full potential for human benefit.

AI encompasses a range of technologies enabling machines to perform tasks that traditionally required human intelligence.

The field includes various methodologies such as machine learning, deep learning, natural language processing, and computer vision.

AI systems derive efficiency from big data and computational power, mimicking human cognitive functions.

The development of AI raises ethical concerns around bias, privacy, transparency, and accountability.

Effective governance of AI requires a multidisciplinary approach involving computer science, law, ethics, and social sciences.

In conclusion, AI is a multifaceted and evolving field that holds tremendous potential and poses significant ethical considerations.

Understanding AI's foundational principles provides the necessary backdrop for exploring the ethical challenges it presents. As technology continues to advance, so too must our efforts to ensure responsible and equitable AI development. By prioritizing ethics at every stage, we can leverage AI in ways that enhance societal well-being while minimizing risks.

Early Ethical Considerations

Artificial Intelligence (AI) has captivated humanity's imagination for decades, promising new horizons and unparalleled opportunities. But with great power comes the need for responsible stewardship. As developers, policymakers, researchers, and tech enthusiasts, the pursuit of technological advancement must be tempered with careful ethical reflection. Early ethical considerations in the realm of AI laid the groundwork for the principles and frameworks that govern its development and deployment today.

The origins of AI ethics can be traced back to philosophical inquiries about the nature of intelligence and morality. Questions about the rights and responsibilities of intelligent agents, whether organic or synthetic, have long fascinated thinkers. Early ethical considerations focused heavily on defining what it means for a machine to act or think "ethically." Philosophers pondered whether machines could possess moral agency or if they were merely extensions of human will. These preliminary discussions set the stage for more structured and rigorous explorations in AI ethics.

One of the most significant early ethical concerns was the potential impact of AI on employment. Even before AI achieved the capabilities it has today, foresighted thinkers worried about machines replacing human labor. The Luddites of the 19th century, who destroyed textile machinery fearing they would lose their jobs, echo contemporary fears about automation. Industries across the board face the prospect of robotic efficiency replacing human jobs, raising urgent ethical questions about economic displacement and the future of work.

Security and safety also emerged as early ethical considerations. What happens when an intelligent system goes rogue or gets hacked? The potential for misuse of AI in warfare, cyber-attacks, or even domestic policing presents real and tangible threats. The dual-use nature of AI technology—meaning it can be used for both beneficial and harmful purposes—requires stringent ethical scrutiny from its inception. The foresight into preventing harmful use cases is as crucial as promoting beneficial ones.

Privacy concerns were another focal point in the early stages of AI development. With the ability of AI to process and analyze massive amounts of data, the risk of infringing on individual privacy became apparent. Early ethical discussions highlighted the imperative to safeguard personal data and to consider the ramifications of data collection practices. This set the stage for current data protection regulations and practices aimed at mitigating privacy risks.

Ethics in AI is not solely about preventing harm; it is equally about fostering fairness and justice. Early discussions often interrogated whether AI could perpetuate or even exacerbate existing biases in society. If a machine learning model is trained on biased data, it can produce biased outcomes, perpetuating systemic inequalities. Ensuring fairness in AI systems is a challenge that demands ongoing ethical vigilance and methodological rigor.

The transparency and explainability of AI systems arose as another crucial ethical consideration. Early on, it became clear that for AI to be trusted and widely adopted, its decisions must be understandable to humans. The "black box" nature of some AI algorithms, where even the developers can't fully explain how decisions are made, poses significant ethical dilemmas. Being able to demystify AI processes ensures accountability and builds public trust.

Accountability, in itself, remains a cornerstone of AI ethics. Early thinkers were concerned with defining who is responsible when an AI system causes harm. Is it the developer, the user, or the machine itself? Establishing clear lines of accountability was, and continues to be, essential for ethical AI deployment. It underpins legal standards and helps in creating ethical guidelines that hold developers and users responsible for their systems.

Autonomy and consent also garnered attention as foundational ethical principles. As AI systems became more capable of making autonomous decisions, it was vital to consider what level of autonomy is ethical. Informed consent became a critical issue, especially in sectors like healthcare and finance, where AI systems would directly impact individuals' lives. Early ethical considerations emphasized ensuring that users fully understand what they consent to when interacting with AI systems.

In sum, the foundational ethical considerations in AI were multifaceted, addressing a broad spectrum of concerns from economic impacts to accountability, fairness, and transparency. The early settlers in the landscape of AI ethics laid the groundwork upon which current ethical frameworks are built. They provided invaluable insights that continue to guide us as we navigate the complex terrain of AI ethics, aiming to harness the power of AI responsibly and ethically.

As we move forward, it is crucial to continually revisit and refine these early ethical considerations. The field of AI is evolving at a rapid pace, and new ethical challenges will undoubtedly emerge. Nonetheless, the fundamental concerns that guided early thinkers will remain as guiding principles, ensuring that AI can be both innovative and ethically sound.

Historical Context

The roots of artificial intelligence (AI) can be traced back to ancient civilizations and their myths, where intelligent automatons and mechanical beings captured human imagination. Legends such as the Greek myth of Talos, a giant, bronze automaton, and the Jewish folklore of the Golem reflect early conceptualizations of artificial beings endowed with intellect and agency. These stories highlighted both the awe and fear associated with creations that could potentially rival human capabilities.

With the dawn of the modern era came scientific inquiry and technological advancement, setting the stage for the eventual birth of AI as a field of study. The concept of mechanized reasoning first gained momentum during the Enlightenment, when philosophers such as Leibniz and Descartes pondered the nature of human intelligence and the possibility of replicating it through machines. Descartes' assertion "Cogito, ergo sum" (I think, therefore I am) marked a significant step in considering thought and reasoning as computational processes that might be mirrored artificially.

Fast forward to the 19th and early 20th centuries: The convergence of ideas from mathematics, engineering, and psychology paved the way for a more concrete exploration of AI. Charles Babbage and Ada Lovelace were at the forefront of this movement, with Babbage's designs for the Analytical Engine and Lovelace's pioneering work on algorithms demonstrating the potential of machines to perform complex calculations and tasks. Although mechanical limitations hindered their success, their theoretical contributions laid foundational stones for future developments.

The field of AI as we recognize it today began to take shape in the mid-20th century. British mathematician and logician Alan Turing is often hailed as a founding figure for his seminal work during this period. Turing's 1950 paper, "Computing Machinery and Intelligence," introduced the iconic Turing Test, a proposed measure of a machine's ability to exhibit intelligent behavior equivalent to, or indistinguishable from, that of a human. This paper posed significant questions about the nature of intelligence and machine consciousness, debates that continue to echo through contemporary AI ethics discussions.

Simultaneously, the development of programmable digital computers during and after World War II provided the necessary hardware for AI experiments. Early pioneers like John von Neumann and Norbert Wiener played key roles in advancing computational theories and cybernetics, emphasizing the interrelation of control and communication in both machines and living beings. Their interdisciplinary work fostered an environment ripe for the nascent AI field to flourish.

The formal establishment of AI as a distinct academic discipline is often marked by the 1956 Dartmouth Conference, organized by John McCarthy, Marvin Minsky, Nathaniel Rochester, and Claude Shannon. This landmark event brought together a diverse group of researchers to discuss the possibility of creating "thinking machines." It's here that McCarthy famously coined the term "artificial intelligence." The conference catalyzed significant interest and investment in AI research, leading to the creation of dedicated AI laboratories and programs at prestigious institutions.

Throughout the 1960s and 1970s, AI research achieved several notable milestones despite encountering formidable challenges. Projects such as the Logic Theorist by Newell and Simon, and the General Problem Solver, endeavored to mimic human problem-solving skills using symbolic reasoning. Meanwhile, Joseph Weizenbaum's ELIZA program demonstrated the potential for natural language processing, albeit in a limited conversational scope. These early successes paradoxically highlighted the field's limitations, laying bare the complexity and unpredictability of human cognition and interaction.

The periods of enthusiasm in AI research often alternated with "AI winters," stretches of reduced funding and interest due to unmet expectations and slow progress. Notably, the AI winter of the late 1970s and early 1980s stemmed from an overestimation of AI's potential combined with underestimation of its complexities. Nonetheless, these downturns spurred deeper introspection and recalibration among researchers, prompting more robust and realistic approaches.

AI began to regain momentum in the mid-1980s with the advent of expert systems, which utilized rule-based approaches to simulate the decision-making abilities of human experts in specific domains. Systems such as MYCIN for medical diagnoses and XCON for computer configuration exemplified practical applications that drove commercial and industrial interest. However, the reliance on manually crafted rules exposed the limitations of scalability and adaptability, issues that would be addressed in later AI paradigms.

The late 20th and early 21st centuries marked transformative periods for AI, driven by advancements in machine learning, neural networks, and data availability. The development of backpropagation algorithms in the 1980s revivified interest in neural networks, which had languished since their inception by McCulloch and Pitts in the 1940s. Subsequent breakthroughs in deep learning, primarily in the 2010s, revolutionized AI, enabling unprecedented capabilities in image recognition, natural language processing, and other complex tasks.

This era also witnessed a surge in ethical considerations surrounding AI. As AI systems began to permeate everyday life, concerns about privacy, bias, transparency, and accountability came to the fore. Historical precedents for ethical thought in technology—spanning medical ethics, bioethics, and information ethics—offered valuable frameworks, yet the unique attributes of AI necessitated novel ethical paradigms and guidelines.

In recent years, multidisciplinary efforts have gained traction, integrating perspectives from the humanities, social sciences, and various technological domains to forge a holistic understanding of AI ethics. Initiatives such as the Asilomar AI Principles and the development of guidelines by the European Commission exemplify proactive steps undertaken by global entities to address and preempt ethical challenges. These endeavors reflect a growing recognition of the profound societal impacts AI holds, underscoring the imperative for responsible design and deployment.

Notably, the historical trajectory of AI ethics is marked by an ongoing dialogue between optimism and caution. While AI promises transformative benefits across sectors, it also poses risks and ethical dilemmas that compel rigorous scrutiny and thoughtful action. By tracing the historical context, we gain insights not only into the evolution of AI but also into the persistent and emergent ethical questions that shape its future.

The landscape of AI ethics is ever-evolving, influenced by technological advancements, societal shifts, and philosophical debates. As we forge ahead, reflecting on the historical context equips us with the wisdom to navigate the ethical complexities of AI with foresight and humanity. In doing so, we stand better positioned to harness the power of AI in ways that align with our deepest values and aspirations.

CHAPTER 2:
PRIVACY AND DATA PROTECTION

As we delve into the realm of privacy and data protection, it becomes clear that these elements are the bedrock of responsible AI development. Our increasingly interconnected world presents both opportunities and risks: while AI technologies can harvest and analyze vast amounts of data for beneficial purposes, they also pose significant threats to individuals' privacy. Understanding the complexities of privacy risks and the ethical implications of data collection practices is essential. It's not just about adhering to the regulations—it's about fostering trust and ensuring that users' data is treated with the utmost respect. Effective strategies to mitigate privacy risks can help navigate this complex landscape. Therefore, adopting proactive measures is paramount to safeguard sensitive information and cultivate a digital environment where ethical AI practices flourish.

Understanding Privacy Risks

As the proliferation of artificial intelligence (AI) continues to weave into the fabric of our daily lives, understanding privacy risks is not just essential; it's imperative. AI systems, by their very nature, require vast amounts of data to function effectively. This data often includes personal information, ranging from text inputs and location data to potentially sensitive details like biometric information. The aggregated collection of such a broad spectrum of data carries inherent privacy risks that need robust scrutiny.

At the heart of these privacy risks is the potential for misuse. Data that is collected for one purpose may be repurposed for another without the explicit consent or even knowledge of the individuals from whom it was collected. This can lead to a breakdown of trust between the public and entities deploying AI technologies. More worryingly, it can pave the way to situations where personal data is exploited to the detriment of the individual—through identity theft, unauthorized surveillance, or discriminatory profiling.

Moreover, privacy risks extend beyond the individual. Collective risks emerge when data from numerous sources is triangulated to reveal broader patterns about specific communities or demographics. These insights, while potentially beneficial in contexts like public health or urban planning, can also be leveraged for less altruistic ends. For instance, targeted advertising can morph into manipulative campaigning or voter suppression efforts, challenging the ethical landscape of our democratic processes.

A significant concern revolves around the opaque nature of many AI systems. Often termed as "black box" systems, these algorithms operate in a manner that is not easily interpretable, not just by the general public, but sometimes even by the developers who created them. This opacity implies that even conscientious organizations might inadvertently propagate privacy infringements simply because they do not fully comprehend the nuances of how their systems process data.

The rapid pace of AI development has outstripped the establishment of comprehensive regulatory frameworks. This regulatory vacuum exacerbates privacy risks as organizations navigate an undefined ethical landscape. In the absence of clear guidelines, the responsibility falls upon developers and stakeholders to self-regulate, posing the question: what ethical guidelines should one adhere to in these uncharted waters?

In addition to the aforementioned risks, there's the growing concern of data breaches. The more data that is collected and stored, the larger the honeypot for malicious actors. High-profile breaches have shown us time and again that no system is infallible. When AI systems are involved, these breaches can have amplified consequences. A leak of not just surface-level data like names and email addresses but deep, inferred insights can cause irreparable harm to individuals.

Consider also the secondary effects of privacy risks: as people grow more aware of the potential misuse of their data, they might become increasingly reticent to share it. This "data hesitancy" could stymie the advancement of AI itself, particularly in areas that rely heavily on user-generated data to refine their algorithms. Therefore, addressing privacy risks doesn't just protect individuals; it ensures the sustainable progress of AI technology.

The global nature of data collection and AI deployment introduces another layer of complexity. Data often traverses international borders, subjecting it to a patchwork of regulatory environments. Regulatory phrases like General Data Protection Regulation (GDPR) in Europe or the California Consumer Privacy Act (CCPA) in the United States outline specific protections, but the inconsistency between these regulations poses challenges. Companies must navigate this patchwork carefully to avoid non-compliance, especially in multinational deployments.

Furthermore, the ethical dilemma is heightened by the prospect of anonymized data. While anonymization is often touted as a method to mitigate privacy risks, the reality is less reassuring. Advanced AI techniques make it increasingly possible to de-anonymize data, thereby re-identifying individuals with a high degree of accuracy. This potential to reverse engineer anonymized data sets underscores the necessity for stringent privacy measures at every stage of data handling.

Privacy risks are multifaceted, with consequences that ripple through social, economic, and political spheres. Developers and stakeholders in the AI ecosystem must prioritize ethical considerations to effectively manage these risks. This isn't merely a technical challenge but a moral imperative, requiring a concerted effort to principles such as data minimization, purpose limitation, and enhanced security practices.

Data minimization calls for the collection and retention of only the data necessary for a specified purpose. Adhering to this principle reduces the amount of information that could potentially be compromised. *Purpose limitation* ensures that data is only used for the purposes explicitly stated at the time of collection, preventing unauthorized or unexpected uses. Enhanced security practices encompass the implementation of robust encryption, regular security audits, and comprehensive incident response strategies to protect data from breaches.

The role of education cannot be overstated in managing privacy risks. Stakeholders, including developers, users, and policymakers, need to be educated on the significance of privacy and the potential risks associated with data misuse. This education forms the bedrock upon which conscientious decision-making rests, allowing for a more informed approach to both the development and application of AI technologies.

Lastly, continuous dialogue between all involved parties is essential. We must foster environments where stakeholders can collaboratively address emerging privacy concerns and establish best practices. This can take the form of public consultations, interdisciplinary forums, and collaborative research initiatives. By engaging in these dialogues, we can stay ahead of potential risks, continually adapt our strategies, and work towards an ethical AI future.

In this intricate dance between technological innovation and ethical responsibility, understanding privacy risks in AI is about looking beyond the immediate. It requires a deep, introspective examination of our values, priorities, and the kind of societal principles we wish to uphold. Only then can we harness the transformative power of AI in a manner that is respectful, equitable, and just.

Data Collection Practices

The collection of data lies at the heart of modern AI systems. Every interaction, click, and movement feeds into vast datasets that fuel machine learning algorithms and predictive models. It's a necessity for creating intelligent systems that can recognize patterns, make decisions, and improve over time. However, the practices surrounding data collection often pose significant ethical concerns that cannot be ignored.

Data is gathered from a multitude of sources - social media platforms, browsing histories, online transactions, and even physical locations through mobile devices. This omnipresence of data collection can often be invisible to the average user. The opacity of these practices can lead to a sense of surveillance and a loss of autonomy, undermining trust in technology. Encouraging transparency in how data is collected, stored, and used is fundamental to maintaining ethical standards.

Data collection practices must adhere to principles of consent and transparency. Users should be aware of what data is being collected, how it will be used, and who will have access to it. While this might seem straightforward, the reality can be murky. Complex terms and conditions often serve more to obscure than to illuminate. Transforming these documents into clear, jargon-free explanations is essential for empowering users.

In addition, companies need to ensure that the data collected is minimized to what is strictly necessary for the intended purpose. Over-collection of data not only increases the risk of privacy breaches but also raises ethical questions about the exploitation of user information. The principle of data minimization encourages establishing limits and protections around data usage.

Furthermore, the provenance of data should be verifiable. With data coming from diverse and often unregulated sources, ensuring the reliability and legality of this data becomes a challenge. Verification processes need to be robust, ensuring that the data fed into AI systems is free from hidden biases, inaccuracies, or illegalities.

Data anonymization and pseudonymization are practices that can significantly mitigate privacy risks. When done properly, these techniques can decouple personal identifiers from data points, protecting individual privacy. However, these methods are not foolproof. Advanced re-identification techniques can sometimes reverse these processes, posing new challenges for data protection.

One of the most critical aspects of data collection practices is the issue of security. Data breaches and leaks can have catastrophic implications, revealing sensitive personal information to unauthorized entities. Consistent and rigorous security measures are required to protect data at all stages - during collection, transmission, and storage. Encryption, access controls, and regular audits are fundamental components of a robust data security strategy.

Another consideration is the ethical sourcing of data. For instance, scraping data from social media without explicit permission from users can be problematic. While the data might be publicly available, its use by AI systems without user consent violates principles of respect and autonomy.

Lastly, the concept of data ownership needs to be addressed. Who owns the data - the individual who generates it, the platform that collects it, or the entity that processes it? This question is not merely academic; it has real-world implications for how data is used and monetized. Clear policies defining data ownership and users' rights over their data should be instituted to ensure fair use and distribution of benefits derived from data.

In many ways, data collection practices can dictate the ethical landscape of AI. By enforcing strict guidelines, maintaining transparency, enhancing security, and respecting user autonomy, we can develop AI systems that are not only intelligent but also ethical and trustworthy.

Mitigating Privacy Risks

Artificial intelligence (AI) holds the potential to transform many facets of society, ranging from healthcare to law enforcement. Yet, while the benefits are substantial, so are the privacy risks. It is crucial to address these risks effectively to ensure that both individuals and society can reap the rewards of AI without compromising personal privacy. Mitigating privacy risks demands a multi-faceted approach that combines technology, policy, and ethical guidelines.

Firstly, one of the fundamental ways to mitigate privacy risks is through data anonymization. This process involves stripping personally identifiable information (PII) from datasets so that individuals cannot be readily identified. By doing so, developers can utilize large volumes of data to train machine learning models without directly infringing on anyone's personal privacy. Techniques such as differential privacy can further enhance data anonymization by adding noise to the data, ensuring that individual records remain confidential even during complex data analyses.

Another key approach is data minimization, a principle that advocates for collecting only the data strictly necessary to achieve a specific objective. This contrasts sharply with current practices where vast amounts of data are often collected "just in case" they might be useful later. By adhering to data minimization principles, organizations can significantly reduce the potential for data breaches and misuse. Algorithms can be designed to require minimal data inputs while still delivering accurate results, thereby striking a balance between utility and privacy.

Consent is another cornerstone in mitigating privacy risks. Users should have clear and transparent information about what data is being collected, how it will be used, and who will have access to it. More importantly, consent should be an ongoing process rather than a one-time formality. Methods like granular consent options allow users to select specific aspects of their data that they agree to share, adding another layer of control over personal information. Implementing easily accessible opt-out mechanisms can also reassure users that their privacy preferences are respected.

On the technological front, encryption plays an essential role in safeguarding data both in transit and at rest. Advanced encryption standards ensure that even if data is intercepted or accessed illegally, it remains incomprehensible without the decryption key. Incorporating end-to-end encryption in AI systems can help protect user data from unauthorized access and tampering, thereby enhancing overall data security.

Furthermore, the concept of federated learning offers a promising solution. Instead of centralizing data for model training, federated learning allows the AI models to be trained on decentralized data located on local devices. This approach not only improves data security by keeping data local but also mitigates the risk of data breaches at a central repository. Federated learning also aligns well with data minimization principles, as it reduces the necessity for massive data transfers.

Legislation and policy also have pivotal roles in mitigating privacy risks. Comprehensive data protection laws like the General Data Protection Regulation (GDPR) in Europe and the California Consumer

Privacy Act (CCPA) in the United States set stringent requirements for data handling practices, providing robust frameworks for protecting individual privacy. These laws mandate transparency, user consent, and the right to be forgotten, ensuring that organizations adhere to ethical data management practices.

Moreover, developing and promoting ethical guidelines for AI research and development is crucial. Frameworks such as the Ethical AI Guidelines provided by various tech consortiums advocate for responsible data practices, prioritizing transparency, accountability, and user autonomy. By adhering to such guidelines, AI practitioners can foster a culture of ethical awareness and responsibility, which is integral to mitigating privacy risks.

In addition to these measures, regular audits and assessments of data protection practices are essential. Routine evaluations help identify potential vulnerabilities in AI systems, enabling timely interventions to mitigate risks. Audits can be conducted by independent third parties to ensure an unbiased assessment, providing additional layers of accountability.

An interdisciplinary approach involving collaboration between technologists, ethicists, policymakers, and other stakeholders is also vital. Diverse perspectives can uncover unique insights and solutions that a single-discipline approach might overlook. By working together, stakeholders can develop more comprehensive and effective strategies to address privacy risks in AI.

Education and awareness among the general public are equally important. Empowering users with knowledge about their data rights and the implications of data sharing can lead to more informed decision-making. Public awareness campaigns, workshops, and educational materials can help demystify AI technologies and their associated privacy risks, fostering a more informed and vigilant user base.

In a broader sense, fostering a culture of ethical innovation within organizations can drive significant improvements in privacy protection. When the importance of privacy is ingrained in organizational values and practices, it influences every stage of the AI development lifecycle. Including privacy considerations right from the design phase ensures that privacy is not an afterthought but a foundational element of AI systems.

Lastly, fostering a global dialogue about privacy standards and best practices can also contribute significantly to mitigating privacy risks. Open communication and collaboration can lead to the development of internationally recognized standards and guidelines, facilitating a unified approach toward data protection. At the same time, it is essential to recognize and respect cultural and legal differences in privacy perceptions across regions and adapt strategies accordingly.

In conclusion, mitigating privacy risks in AI systems is an ongoing effort that demands vigilance, innovation, and collaboration. By employing a combination of technical measures, policy interventions, ethical guidelines, and public education, we can create AI systems that respect individual privacy while harnessing the transformative potential of artificial intelligence. The journey toward mitigating privacy risks is complex, but with a concerted effort from all stakeholders, it is a goal within our reach.

CHAPTER 3:
BIAS AND FAIRNESS

Bias and fairness have become critical concerns in the development and deployment of artificial intelligence systems. As AI increasingly permeates sectors like healthcare, law enforcement, and finance, the risks posed by unintended biases in algorithms can have significant and far-reaching consequences. Addressing bias isn't just a technical issue; it's a moral imperative that demands a nuanced understanding of both the sources and impacts of bias. Strategies for mitigating bias must encompass comprehensive approaches that involve diverse data sets, inclusive team compositions, and continuous monitoring mechanisms. Fairness in AI also requires a balance between competing ethical principles, ensuring that solutions are not just technically sound but ethically robust. This chapter delves into these challenges, offering insights into how developers and policymakers can work together to create systems that genuinely uphold fairness and equity.

Types of Bias in AI

Artificial intelligence, with its promise to revolutionize industries and tasks alike, faces significant ethical scrutiny, particularly regarding bias. Bias in AI can manifest in various ways, often with serious implications for fairness and equity. Understanding these types of bias is crucial for developers, policymakers, and researchers aiming to create fairer AI systems.

Training Data Bias

One of the most talked-about types of bias in AI is training data bias. This occurs when the data used to train an AI model does not represent the diversity of the real world. If an AI system is trained on data that skews toward a particular demographic—be it race, gender, or socioeconomic status—the system will likely produce biased results. For instance, facial recognition technologies have been criticized for higher error rates in identifying individuals from certain racial groups, primarily because the training data lacked diversity. This bias can perpetuate inequality by reinforcing societal stereotypes and potentially discriminating against underrepresented groups.

Another example is language models that perform poorly on dialects or languages less represented in the training set. These inconsistencies can lead to exclusionary technologies that fail to meet the needs of a global population. To mitigate training data bias, it's essential for datasets to be as comprehensive and representative as possible.

Algorithmic Bias

Algorithmic bias is another form that can occur even when the training data is unbiased or comprehensive. This type arises from biases in the design and implementation of the algorithm itself. For instance, the objective functions, heuristics, or assumptions made during development can introduce bias, often unintentionally. If an algorithm prioritizes efficiency over fairness, it might end up disadvantaging certain groups. The issue of algorithmic bias isn't just technical; it's deeply ethical, calling for transparent and inclusive processes during the development phase.

Moreover, different algorithms might interact in unforeseen ways, compounding existing biases. This complexity necessitates robust testing and auditing frameworks, where algorithms are continuously assessed for unintended consequences. Establishing interdisciplinary teams can help ensure that diverse viewpoints are considered, reducing the likelihood of inherent biases.

Human Bias

Humans, who design, develop, and maintain AI systems, bring their own biases into the process. This human bias can be explicit or implicit. For instance, if a developer harbors unconscious biases, these can seep into the AI's decision-making framework. Examples include gender biases in hiring algorithms or racial biases in predictive policing tools. Human biases are challenging to eliminate but recognizing their existence is the first step. Awareness programs and bias training for developers and stakeholders can go a long way in mitigating this form of bias.

Institutions should foster an environment that values diversity, encouraging contributions from people of various backgrounds. This diversity can act as a check against the homogeneous viewpoints that often perpetuate existing biases. Including ethicists and social scientists in the AI development process can provide a critical lens through which algorithms are evaluated.

Measurement Bias

Measurement bias occurs when the metrics or variables used to inform an AI model are inherently biased. For instance, if an AI system uses past arrest records to predict future criminal activity, it might overlook systemic issues like racial profiling and over-policing in certain communities. Since the measurements themselves are flawed, the AI's predictions will also be biased. Addressing this requires rethinking the metrics and variables used in AI models to ensure they are equitable and accurate.

The selection of features in an AI system must be scrutinized rigorously. Simplistic or inappropriate metrics can lead to skewed outcomes. This calls for comprehensive validation processes that take into account the socio-technical contexts within which these systems operate.

Preconceived Bias

Preconceived bias is closely related to human bias but focuses more on the goals and perspectives of those who commission or use AI systems. For example, if a company directs an AI system to maximize profits at all costs, the AI might adopt unscrupulous practices like microtargeting vulnerable consumers. Here, the bias stems not only from the developers but also from the stakeholders setting the system's objectives.

This type of bias can be counteracted by establishing ethical guidelines that govern the goals and uses of AI technologies. Policies should encourage the alignment of AI objectives with broader societal values, ensuring that the technology works for the common good rather than specific vested interests.

Exclusion Bias

Finally, exclusion bias occurs when certain groups are systematically left out of the AI system's purview. For instance, healthcare algorithms often perform poorly on underrepresented populations, exacerbating existing disparities in healthcare access and quality. Exclusion bias can be unintentional but has severe consequences for affected groups.

The remedy lies in inclusive research and development practices. Consulting with affected communities, conducting impact assessments, and iteratively updating models to include previously excluded groups are steps in the right direction. Collaborative efforts between stakeholders, including governmental and non-governmental organizations, can help ensure inclusivity.

In summary, understanding the various types of bias in AI—training data bias, algorithmic bias, human bias, measurement bias, preconceived bias, and exclusion bias—is essential for building fairer systems. Each type of bias demands its unique approach for mitigation, from diversifying training data and fostering interdisciplinary development teams to critically reassessing metrics and ensuring inclusive practices. Addressing these biases is not just a technical challenge; it's a societal imperative that requires ongoing vigilance and a commitment to ethical principles. Through collective effort, we can strive towards AI systems that benefit all of humanity without perpetuating existing inequities.

Detecting Bias

Detecting bias in artificial intelligence (AI) systems is an essential step toward ensuring fairness and ethical integrity. As AI becomes more integrated into various aspects of society, from employment to healthcare, recognizing and addressing bias within these systems is crucial. Bias can emerge in multiple forms, whether it's through data, algorithms, or user interactions, each contributing to the potential for discrimination and unfair outcomes. To mitigate these risks, a multifaceted approach is necessary to identify and rectify bias at every stage of an AI system's lifecycle.

One of the primary methods for detecting bias involves thorough data analysis. Data is the foundation of AI, and biases inherent in the data can influence the outcomes of any AI model. By scrutinizing datasets for imbalances or anomalies, we can identify potential sources of bias. This analysis often involves statistical checks to ensure demographic groups are fairly represented. Techniques such as re-sampling, stratification, and the use of domain expertise are employed to uncover hidden patterns that may indicate bias. However, data analysis alone isn't sufficient; it must be complemented by other measures.

An essential tool in the detection of bias is the development and use of fairness metrics. These metrics provide quantifiable means to evaluate whether an AI system is performing equitably. Common metrics include disparate impact ratio, equalized odds, and demographic parity, among others. Each of these metrics offers different insights into how and where biases may manifest in AI outputs. For instance, disparate impact measures the differential effects of a decision on various demographic groups, revealing whether certain groups are disadvantaged by specific outcomes. Implementing these metrics requires a deep understanding of the AI model and the context in which it operates.

Beyond data and fairness metrics, testing AI systems in varied and controlled environments is another critical step. Controlled testing involves creating scenarios that mimic real-world applications but in a controlled manner to observe how AI systems make decisions. By exposing AI to diverse inputs and monitoring the results, it's possible to identify discriminatory patterns that might not be evident during the initial training phases. This form of testing also allows developers to simulate edge cases—rare or unusual scenarios that the AI might encounter—providing further insight into how bias can emerge and affect decision-making processes.

One significant challenge in detecting bias is the phenomenon known as feedback loops. Feedback loops occur when biased outcomes reinforce the data that trains future iterations of the AI model, perpetuating and even amplifying the original bias. It's crucial to monitor AI systems over time, not just during their initial deployment but throughout their usage lifecycle. Continuous monitoring and periodic re-evaluation of the system's outputs can help detect bias early and prevent feedback loops from escalating. Implementing automated monitoring tools can aid in this by providing real-time alerts when unusual patterns or discrepancies are detected.

Collaborative approaches also enhance the process of bias detection. Engaging a diverse group of stakeholders, including domain experts, ethicists, and representatives from affected communities, can provide invaluable perspectives and insights. These stakeholders can identify biases that might not be apparent to developers or data scientists who are too close to the problem. Collaborative frameworks can also help establish trust and ensure that the AI system is considered fair and legitimate by various societal groups. Such engagement should be ongoing, creating a dynamic dialogue that continually refines and improves bias detection strategies.

Furthermore, leveraging transparency and explainability is critical in bias detection. Explainable AI (XAI) techniques seek to make the decision-making processes of AI systems more understandable to humans. By elucidating how an AI reaches its conclusions, we can better identify where biases may be occurring. Transparency helps build accountability, allowing stakeholders to question and audit the AI's behavior. Several approaches to XAI exist, from rule-based systems and interpretable models to more advanced techniques like Layer-wise Relevance Propagation (LRP) and Shapley values. Each of these methods offers different levels of insight into the AI's reasoning, enabling targeted investigations into potential biases.

AI audits and third-party evaluations are also effective tools for detecting bias. Independent audits by external organizations can provide an objective assessment of an AI system's fairness. These audits typically involve a comprehensive review of the system's development process, data sources, and performance metrics. Third-party evaluations can uncover biases that internal teams might overlook, offering a fresh perspective on how the AI functions in practice. These evaluations not only contribute to bias detection but also enhance the credibility and trustworthiness of the AI system among users and stakeholders.

Finally, regulatory frameworks and standards play a pivotal role in guiding the detection of bias. Emerging regulations worldwide are beginning to mandate bias detection and mitigation as part of compliance requirements for AI systems. These regulatory guidelines serve as a baseline for acceptable practices, pushing organizations to adopt more rigorous methods for identifying and addressing bias. While regulations can set minimum standards, industry leaders must often go beyond these requirements to truly achieve fair and unbiased AI.

To sum up, detecting bias in AI involves an intricate blend of data analysis, fairness metrics, controlled testing, continuous monitoring, stakeholder collaboration, transparency practices, third-party audits, and regulatory compliance. Each of these elements contributes to a holistic strategy that aims to identify and mitigate bias at all levels of an AI system. The stakes are high, as biases in AI can lead to significant ethical and social consequences, reinforcing existing inequalities and undermining trust in technology. Therefore, it's imperative for developers, policymakers, researchers, and tech enthusiasts alike to remain vigilant and proactive in the quest for unbiased AI systems. The journey toward fair and ethical AI is ongoing, demanding a continuous commitment to detection, understanding, and rectification of biases, as we strive to create technology that serves all of humanity equitably.

Strategies for Reducing Bias

Bias in artificial intelligence isn't just an abstract problem; it has real-world implications that can affect decisions in criminal justice, healthcare, hiring, and beyond. Strategies for reducing bias involve meticulous planning, continuous monitoring, and an unwavering commitment to fairness. Here, we'll explore various approaches that can be implemented to mitigate bias at different stages of AI development and deployment.

One of the most fundamental strategies is diversifying data sets. Often, biased decisions stem from training AI models on data that isn't representative of the population it's meant to serve. By actively seeking out underrepresented groups and including their data, we can strive to create more balanced and equitable systems. This isn't merely about quantity but also quality; ensuring the data collected from diverse sources is accurate and reliable is crucial. For example, if an AI system is trained to recognize human faces, collecting data from a wide range of ethnicities, ages, and other demographic groups can help in reducing error rates and biases.

Another pivotal strategy is algorithmic transparency. When users and developers can understand how an algorithm makes its decisions, it's easier to spot and address biases. This involves documenting design choices, providing access to key metrics, and explaining the rationale behind each decision the AI makes. Transparency doesn't just build trust; it's a practical step towards accountability. When AI systems are black boxes, biases can embed themselves unnoticed. Opening these boxes to scrutiny helps us identify biases we might not even be aware of.

Regular auditing and monitoring of AI systems is another powerful tool. Biases can emerge over time, especially as societal norms and datasets evolve. Regular audits can include checking for disparate impacts on different groups, running tests with edge cases, and reviewing decision-making processes periodically. Audits should be conducted by independent teams to avoid conflicts of interest. Continual monitoring ensures that the AI adapts to new data and evolving societal values without perpetuating outdated biases.

Human oversight is indispensable in the quest to reduce bias. AI systems should be designed to include human intervention points where necessary. This might involve humans reviewing decisions made by the AI, particularly in high-stakes situations like parole decisions or medical diagnoses. Incorporating human judgment helps to balance out the potential shortcomings of an algorithm and provides a safety net against erroneous decisions.

Bias in AI can also be mitigated through collaborative approaches. Developers can't operate in a vacuum if they aim for fairness and equity. Engaging with ethicists, sociologists, and representatives of affected communities during the development process can provide valuable insights that might otherwise be overlooked. This multidisciplinary approach helps to address potential biases from multiple angles and fosters a culture of inclusiveness.

Education and training for AI developers and users form the bedrock of long-term bias reduction. Awareness of the different types of biases and understanding their sources can empower individuals to make more conscious decisions. Workshops, courses, and continuous learning programs focusing on ethics and fairness can prepare developers to tackle bias proactively. Knowledge dissemination isn't just for the creators; educating end-users can also help in identifying and reporting biases, making the entire AI ecosystem more robust.

Technical measures such as fairness-aware machine learning algorithms can also play a significant role. These algorithms are designed to minimize discrimination by incorporating fairness constraints directly into the model-building process. For instance, re-weighting data points or adjusting thresholds for different groups can help ensure a more equitable distribution of outcomes. These technical interventions should be part of a comprehensive strategy to address bias, complementing other non-technical measures like policy changes and ethical guidelines.

Implementing feedback loops is another crucial strategy. AI systems should be capable of learning from their mistakes and improving over time. This involves setting up mechanisms where users can flag biased decisions, and developers can then use this feedback to adjust algorithms. Feedback loops create a dynamic and evolving system that can adapt to new types of bias as they emerge, ensuring long-term fairness and reliability.

Moreover, regulation and policy can provide the framework needed to enforce fairness in AI. Governments and international bodies can set standards and guidelines that mandate the inclusion of fairness metrics in AI development. Policies that require transparency reports, bias audits, and fair data practices can create an environment where ethical AI is not just encouraged but required. Regulatory frameworks can also aid in the standardization of fairness practices across different industries and regions.

Public engagement is another underrated yet powerful strategy for reducing bias. By involving the general populace in discussions about AI fairness, we can democratize the understanding and impact of these technologies. Public forums, community consultations, and transparent reporting practices can help demystify AI and make its operations more answerable to the people it serves. Public scrutiny and feedback can provide a diverse array of perspectives, helping to identify biases that the developers might have missed.

Lastly, fostering a corporate culture that prioritizes ethics and fairness is indispensable. Organizations should integrate principles of fairness into their core values and operational strategies. Encouraging ethical behavior, rewarding transparency, and penalizing unethical practices can create an environment where fairness naturally thrives. Leadership should take an active role in promoting these values, ensuring that every level of the organization is aligned in the quest for unbiased AI.

While these strategies collectively aim to reduce bias, it's crucial to realize that no single approach is sufficient on its own. The multifaceted nature of bias requires a comprehensive, persistent, and adaptive effort. Reducing bias is not a one-time task but an ongoing process that evolves as society, data, and technology change.

CHAPTER 4:
TRANSPARENCY AND EXPLAINABILITY

In today's rapidly-evolving AI landscape, transparency and explainability stand as pillars essential to fostering trust, enhancing accountability, and ensuring ethical practices. Unlike opaque "black box" systems, which obscure the decision-making processes and algorithms, transparent AI models demystify how they function and upon what data they act. This clarity not only empowers users and stakeholders to understand AI actions but also enables developers to identify and rectify potential biases or errors in the system. When AI systems provide comprehensible explanations for their decisions, it democratizes access to technology, making it more accessible to those without specialized knowledge. Moreover, transparent and explainable AI bolsters informed policy-making, helping legislators craft laws that protect privacy and prevent misuse. By advocating for and integrating transparency and explainability, we cultivate a future where AI operates in a manner aligned with societal values, ethical principles, and human rights.

Why Transparency Matters

Transparency stands as one of the ethical pillars supporting responsible AI systems. It's more than a mere ideal; it's a necessity for ensuring trust, accountability, and fairness. Transparent systems allow users, developers, and policymakers to gain insights into the mechanisms driving AI decisions. Without transparency, we're navigating a maze of complexities and uncertainties, making it nearly impossible to hold systems and their creators accountable. Think of it as removing the cloak from a magician's performance—only then do we understand the mechanics behind the illusions.

Consider the real-world impact of non-transparent AI. When facial recognition systems are employed in policing, the stakes are incredibly high. If these systems operate without transparency, affected individuals can't scrutinize or challenge the decisions made about them. This lack of visibility shifts power imbalances further, leaving vulnerable populations at greater risk for discrimination and errors. Transparency provides a pathway for societal checks and balances, ensuring that technology serves humans equitably and justly.

A transparent AI system is like a well-lit classroom where everyone, from the teacher to the students, can follow the lesson and understand when an error occurs. This openness not only demystifies the system's operations but also instills a sense of trust. Users are more likely to accept and engage with AI when they can see and understand its logic and workings. It's a form of intellectual honesty that fosters collaborative growth and refinement of the technology.

The complexity of AI algorithms often means that they are enigmatic even to their developers. Black-box algorithms can perform remarkably well in tasks like language translation, image recognition, or predictive analytics, but their opacity can be a double-edged sword. It is not just about knowing the input and output; it's about understanding the process. When errors occur, transparency is the bridge to identifying and rectifying those mistakes. Without it, errors may go unaddressed, perpetuating harm and misuse.

Moreover, transparency works hand-in-hand with explainability, another critical facet of ethical AI. While transparency allows us to see into the system, explainability helps us understand how conclusions are reached. These concepts are intertwined and mutually reinforcing. As we delve deeper into AI's role in sectors like healthcare, law enforcement, and finance, the need for transparent and explainable models becomes all the more pressing. In these areas, erroneous or biased decisions can have significant, far-reaching consequences.

For instance, in healthcare, AI diagnostics can significantly enhance patient care. But a transparent system ensures that healthcare professionals can trust and verify AI recommendations. This layer of scrutiny is vital for preventing misdiagnosis and ensuring patient safety. Similarly, in law enforcement, transparency can act as a safeguard against misuse of AI for surveillance or profiling, allowing for greater accountability and ethical alignment.

The ethical considerations surrounding transparency extend to the socially constructed biases that AI can inherit from historical data. A lack of transparency conceals these biases, making it challenging to identify and correct them. Transparent AI can facilitate the detection and mitigation of biases, contributing to fairer and more equitable outcomes. This is crucial in creating systems that reflect and uphold societal values and norms.

Transparency also fosters innovation and improvement. When AI models are open for scrutiny, the wider community, including researchers and developers, can contribute to their refinement. This collective effort leads to more robust, secure, and effective AI systems. It's akin to the principles driving open-source software, where collaborative scrutiny results in more resilient and versatile programs.

While transparency is polemical for privacy concerns, especially regarding data used for training AI, achieving a balance is feasible. It's possible to design systems that offer insight into their workings without compromising user privacy. Techniques like differential privacy and federated learning are emerging as solutions that allow for transparency without jeopardizing sensitive information. This harmonious coexistence of transparency and privacy is an area ripe for exploration and implementation.

The role of policymakers in promoting transparency cannot be understated. Regulatory frameworks can mandate transparency requirements for AI applications, especially those deployed in critical sectors. Such regulations ensure that organizations follow standardized practices for disclosure, thereby promoting an ecosystem where transparency becomes the norm rather than the exception. This legislative backing provides the necessary impetus for organizations to prioritize ethical considerations in their AI development processes.

But policy alone is not sufficient. The ethical commitment to transparency needs to be an intrinsic part of organizational culture. Tech companies and developers need to prioritize ethical training and awareness among their teams. When the ethos of transparency permeates every level of development, from initial design to deployment and maintenance, the resultant AI systems are inherently more trustworthy and responsible.

Transparency also serves as a foundation for meaningful user consent and autonomy. Users need to understand how their data is being utilized and how AI-driven decisions impact them personally. Transparent systems can provide users with the necessary context to make informed decisions about their interactions with AI technologies. This empowerment fosters a more engaged and informed user base, which is fundamental for the responsible evolution of AI.

In conclusion, transparency is not just a technological or procedural necessity; it is a moral imperative. It serves as the bedrock for trust, accountability, and fairness in AI systems. Transparent AI promotes societal good by allowing public scrutiny, enabling error correction, reducing biases, fostering innovation, and ensuring that the technology aligns with ethical standards and societal values. As AI continues to evolve and permeate various aspects of our lives, the call for transparency becomes louder and more urgent. It is not merely a best practice but an essential requirement for ethical AI advancement. Embracing transparency ensures that AI serves as a force for good, guiding us toward a more equitable and just future.

Explainable AI Models

The quest for explainable AI models has become a focal point of discussions surrounding the transparency and accountability of artificial intelligence systems. These models aim to make AI decisions understandable to humans, thereby fostering trust and ensuring ethical alignment in their operations. Let's delve into what makes AI explainable and why this aspect is crucial in today's AI-driven world.

Explainable AI (XAI) models serve as a bridge between the complex algorithms that drive artificial intelligence and the human need for understanding how decisions are made. They enable users to see beyond the opaque "black box" nature of many AI systems, unveiling the logic or reasoning behind specific outcomes. These models seek to answer questions like "Why was this decision made?" and "What factors influenced this outcome?" Such transparency is critical not just for effective human-AI collaboration, but also for regulatory compliance and ethical responsibility.

One of the primary motivations for developing explainable AI models is to align AI outcomes with societal values and ethical standards. For instance, when AI algorithms are employed in the judiciary system for predictive policing or sentencing, the stakes are incredibly high. The decisions made by these systems can significantly impact individuals' lives. Therefore, it is essential that there is a clear, understandable rationale behind each decision, ensuring that it is fair, unbiased, and justifiable.

Moreover, explainability fosters accountability. When AI models are transparent, it's easier to identify the sources of errors or biases. It provides a means to audit and improve the systems continuously. This accountability is not only beneficial for developers and users but is also a requirement by regulatory bodies in many jurisdictions. Regulations such as the European General Data Protection Regulation (GDPR) mandate that individuals have the right to an explanation of decisions made by automated systems, particularly when those decisions have significant effects.

From a technical perspective, achieving explainability can be notably challenging. Traditional machine learning models like linear regression or decision trees are inherently more interpretable. However, more complex models, such as deep neural networks, present significant hurdles to explainability due to their intricate architectures and large number of parameters. Researchers are constantly developing new methods to interpret these 'black box' models, such as Local Interpretable Model-Agnostic Explanations (LIME) and Shapley values, which seek to provide insights into the functioning of these complex systems.

Incorporating explainability into AI models does not necessarily mean sacrificing performance. While it was once believed that there was a trade-off between model accuracy and interpretability, advances in the field are showing that it is possible to achieve both. Hybrid models that combine interpretable models with less interpretable components, or feature importance techniques that highlight which input features are most influential, are examples of how the balance can be struck.

Aside from technical methods, involving interdisciplinary collaboration is key to creating truly explainable AI. Ethicists, sociologists, and legal experts should work alongside data scientists and engineers to ensure that the AI systems designed align with human values and legal frameworks. This collaborative approach ensures that diverse perspectives are considered and that the AI models developed are both functionally effective and socially responsible.

In practical applications, explainable AI models have shown significant promise. In the healthcare sector, for example, explainable AI can help clinicians understand the reasoning behind diagnostic predictions or treatment recommendations, which can then be communicated to patients in a comprehensible manner. This not only aids in making informed clinical decisions but also enhances patient trust in AI-assisted medical procedures.

Another critical area where explainable AI has shown impact is finance. Financial institutions use AI for credit scoring, fraud detection, and investment decisions. Explainable AI models help these institutions understand the factors leading to these decisions, ensuring that they comply with regulatory requirements and maintain customer trust. By providing clear justifications for actions taken, these models reduce the risk of unfair treatment and enhance the clarity of operational processes.

Education is yet another field benefiting from explainable AI. AI systems can offer personalized learning experiences by adapting to each student's needs. However, for educators and parents to trust and adopt these technologies, understanding how recommendations and adaptations are generated is crucial. Explainability helps stakeholders understand and validate the system's recommendations, ensuring they are in the best interest of the students.

Nevertheless, there are inherent challenges and limitations to be addressed in achieving explainable AI. One challenge is the potential complexity and expense involved in making highly sophisticated models transparent. Often, creating interpretable models involves intricate methodologies that demand significant time and resources, posing a hindrance to their widespread adoption. Additionally, the concept of explainability itself can vary significantly depending on the user's background, making it difficult to create a one-size-fits-all solution.

Moreover, there is a delicate balance to be maintained between preserving the intellectual property and proprietary algorithms of AI developers and providing sufficient transparency to users. Ensuring explainability without revealing proprietary details requires innovative approaches and a commitment to ethical transparency from organizations. Policymakers may need to develop frameworks that strike this balance, ensuring that developers can protect their IP while users receive the necessary explanations.

Another important consideration is the level of explainability needed for different applications. Not all AI applications require the same depth of transparency. For example, a recommendation algorithm for a streaming service might not necessitate the same level of detailed explanation as an AI system used in legal sentencing. Defining appropriate levels of explainability based on context and use-case is essential for the practical implementation of XAI.

As we move forward, the future of explainable AI models lies in continuous development and adaptation. Researchers and practitioners must stay attuned to emerging challenges and advancements, ensuring that AI systems remain transparent and aligned with ethical norms. Collaborative efforts across disciplines and sectors are paramount to developing robust, explainable AI models that serve society's best interests.

Innovation in explainable AI models offers an inspiring path forward in the realm of ethical AI. By continuing to prioritize transparency and accountability, we pave the way for AI systems that not only perform effectively but also command trust and reliability. These systems will be instrumental in a future where AI acts as a partner to humanity, fostering progress across diverse sectors while upholding the highest ethical standards.

In conclusion, explainable AI models represent a critical component in the ongoing effort to ensure ethical and transparent AI systems. They bridge the gap between complex technologies and human understanding, fostering trust, accountability, and ethical alignment in AI-driven decision-making processes. By embracing explainability, we can create AI systems that are not only powerful and efficient but also transparent, responsible, and aligned with societal values.

Communicating AI Decisions

In the complex world of artificial intelligence, transparency and explainability aren't just buzzwords; they're foundational principles that guide ethical AI development. At the heart of these principles lies the crucial need for effectively communicating AI decisions. Without clear communication, even the most transparent and explainable systems fail to build trust and ensure accountability.

The first challenge in communicating AI decisions is breaking down technical jargon into layperson's terms. AI models, especially advanced ones, often involve intricate mathematical constructs and highly technical language. When an end user receives a decision or recommendation from an AI system, they need a translation of the 'black box' into understandable insights. This doesn't mean dumbing down the information but rather contextualizing it in a way that's relevant and accessible.

Imagine you're an independent auditor reviewing an AI-driven loan approval system. You'd want to know why certain applications were approved and others rejected. While the raw output of the model might include logistic regression coefficients or neural network parameters, these mean little without context. Effective communication might instead highlight that 'applicant's credit score' and 'income stability' were weighted heavily in the decision. This kind of transparency fosters trust and allows for fair scrutiny.

Another key aspect is the decision-making timeline. AI decisions often hinge on data that update frequently. For example, ride-sharing apps rely on real-time traffic conditions to set prices dynamically. Users should be aware of how often the data is updated and what time span the data covers. Explaining this process requires not just providing figures but painting a picture of how dynamic inputs influence dynamic outputs.

Variety in communication formats can also significantly enhance understanding. Detailed reports and summaries can cater to different audiences, from technical experts to policymakers. Visual aids such as graphs and flowcharts are incredibly effective at breaking down complex concepts. Interactive dashboards that allow users to manipulate variables and see how decisions are affected can provide another layer of transparency.

This brings us to the role of feedback loops. Communication should not be one-way traffic. End users, auditors, and even the general public must be able to provide feedback on AI decisions. This involves not only listening but acting on feedback to refine explanations and improve decision-making processes. In medical AI, for example, misdiagnoses or patient feedback can help recalibrate machine learning models to capture more nuanced realities.

Language is another critical element. In multilingual societies or global applications, explanations need to be available in multiple languages and cultural contexts. What makes sense in an English language framework might be less clear or culturally inappropriate in another language. Hence, translation isn't merely a matter of different words but capturing and conveying the same essence and accuracy across diverse backgrounds.

The importance of explaining AI decisions extends beyond individual end-users to institutions and regulatory bodies. Organizations deploying AI at scale must ensure regulators understand how their systems work. This not only builds institutional trust but also paves the way for informed policymaking. Comprehensive reports that outline methodologies, data sources, and risk factors can be invaluable tools for both internal and external stakeholders.

To effectively communicate AI decisions, institutions need specialized roles—AI ethicists, transparency officers, and communication experts who bridge the gap between developers and end users. These specialists work in tandem with developers to continuously refine explanatory models and ensure that communication remains clear and effective.

Transparency tools and platforms are proving indispensable in this aspect as well. Explainability frameworks such as LIME (Local Interpretable Model-agnostic Explanations) and SHAP (SHapley Additive exPlanations) visualize the influence of each feature on the outcome of AI models. While these tools are inherent to some degree of technical complexity, they're indispensable for providing a layer of understandable interpretation that can be passed down the communication chain.

Furthermore, ethical guidelines stipulate that users should be informed about the presence and nature of automated decision-making systems. For instance, when interacting with customer service chatbots, users need to be aware they're conversing with AI and understand the parameters guiding the conversation. Transparency about these guidelines helps maintain honesty and builds a rapport of trust.

It's also important to address scenarios where AI systems make errors or adversarial decisions. Communicating these errors, understanding their source, and providing a path for resolution or appeal are crucial for accountability. This entails not just identifying that an error happened but explaining why it happened and what steps are being taken to prevent similar errors in the future. For legal proceedings, such explanations could form the basis for understanding liability and providing remedies.

Machine learning models and AI systems are constantly evolving. Continuous updates and retraining might mean that the basis for decisions changes over time. Therefore, organizations must have mechanisms in place to update their communication in tandem with these changes. A proactive approach ensures that all stakeholders are informed about how new model training data or updated algorithms impact decisions.

One can't overemphasize the importance of regular audits. Independent audits provide an external review that can highlight gaps in communication and offer standardized methods for reporting AI decisions. Transparent auditing processes not only bolster internal practices but also present a front of accountability to the public.

The responsibility for effective communication doesn't lie solely with data scientists or AI developers. Cross-disciplinary teams that include sociologists, psychologists, and communication experts can provide a more holistic approach to understanding how best to convey complex AI decisions. These cross-disciplinary insights can reveal blind spots that purely technical viewpoints might miss.

Lastly, ethical considerations in AI communication transcend language and technical barriers. Trust isn't built in a vacuum; it's a two-way street that requires ongoing effort and commitment. As AI continues to permeate various facets of society, ensuring that its decisions are understandable, transparent, and accountable remains a cornerstone of ethical AI development.

In conclusion, communicating AI decisions isn't just a technical challenge but a societal obligation. It encompasses converting technical outputs into relatable narratives, utilizing a variety of communication formats, incorporating feedback loops, and ensuring multilingual and culturally appropriate dissemination. Only through committed and ongoing efforts in transparent communication can we hope to build

AI systems that are not only efficient but also ethically sound and socially responsible.

CHAPTER 5:
ACCOUNTABILITY IN AI SYSTEMS

In the ever-evolving landscape of artificial intelligence, accountability stands as a crucial pillar ensuring that AI systems operate with integrity and responsibility. When AI technologies fail or cause harm, the question of who is held accountable becomes paramount. Legal and ethical standards guide us, yet the rapid pace of AI innovation often outstrips existing frameworks. Effective accountability mechanisms encompass not only a clear delineation of responsibility among developers, users, and policymakers but also necessitate robust auditing and oversight processes. By fostering a culture of accountability, we empower stakeholders to address and rectify errors, biases, and unintended consequences in AI systems. Ultimately, achieving accountability in AI isn't just about assigning blame; it's about building trust and fostering a future where AI enriches society while upholding ethical principles.

Defining Accountability

Accountability in AI systems serves as a cornerstone in the ethical framework guiding their development and deployment. Yet, accountability is far more complex than merely acknowledging responsibility. It encompasses a spectrum of duties, including the delineation of roles, the clear communication of responsibilities, and the establishment of transparent processes for auditing and rectifying AI decisions. By defining accountability in this multifaceted manner, we set the stage for crafting AI systems that not only perform effectively but also adhere to ethical norms and societal expectations.

51

At its core, accountability means being answerable for actions and decisions. This concept translates into AI systems where accountability must be clearly mapped out across different stakeholders. Developers, policymakers, end-users, and even the AI systems themselves hold varying degrees of responsibility. For instance, developers are accountable for ensuring the code they write adheres to ethical guidelines, while policymakers must create frameworks that support ethical AI practices. Collectively, these layers of accountability create a robust structure designed to minimize harm and maximize benefits.

Furthermore, the challenge of defining accountability in AI is exacerbated by the technical complexity and opacity often inherent in these systems. Unlike traditional software, AI can make decisions that are not explicitly programmed, leading to unpredictable outcomes. This unpredictability necessitates a more rigorous approach to accountability, which includes continuous monitoring and evaluation. Development teams must prioritize creating mechanisms for tracing decision pathways and understanding how an AI system arrives at specific conclusions. Without this transparency, it becomes nearly impossible to hold any party accountable effectively.

Another vital aspect is the ethical dimension of accountability. AI systems can significantly impact individuals and communities, necessitating ethical considerations at every step of their lifecycle. Ethical accountability ensures that systems are designed with human values in mind, avoiding biases that can lead to unfair treatment. For example, an AI used in hiring processes must be scrutinized for any algorithmic biases that could disadvantage particular groups. Ethical accountability, therefore, demands a conscientious effort to align AI operations with broader societal ideals of fairness and justice.

Legal frameworks provide another crucial layer in defining accountability. Unlike ethical guidelines, which are often voluntary, legal requirements have enforceability backed by judicial systems. Various countries and international bodies are actively working on establishing standards and regulations to govern AI development. These legal mandates serve multiple functions: they provide guidelines for responsible AI design, set penalties for violations, and establish formal routes for grievance redressal. The interaction between ethical guidelines and legal requirements forms a comprehensive accountability framework that helps ensure AI systems act in society's best interest.

Accountability also extends to the domain of user interactions. Transparency in user interfaces and clear communication about how AI systems operate empower users to make informed decisions. When users understand the limitations and capabilities of an AI system, they can use it more effectively and responsibly. Consequently, developers need to implement user-centric design principles that prioritize transparency and usability. Enabling users to report failures and issues further fortifies the accountability loop, providing valuable feedback for continuous improvement.

The socio-technical aspect of AI accountability encompasses the broader social and cultural impacts of AI systems. Universities and research institutions play an essential role here by fostering a culture of ethical inquiry and responsible innovation. Academic programs that emphasize ethics in technology, interdisciplinary research collaborations, and public engagement initiatives contribute to a more accountable AI ecosystem. By integrating socio-technical perspectives into the definition of accountability, we ensure that AI benefits society in a holistic manner.

Regulatory bodies face unique challenges in defining and enforcing accountability in AI. Unlike traditional industries where roles are more clearly defined, the AI landscape is incredibly dynamic, necessitating adaptive and forward-thinking regulatory approaches. Regulatory agencies must engage in continuous dialogue with technologists, ethicists, and the public to stay abreast of emerging trends and potential issues. Through this collaborative approach, regulatory frameworks can be more responsive and effective in governing AI systems.

Another consideration in defining accountability is the need for ethical audits and certifications. Third-party audits provide an objective assessment of an AI system's compliance with ethical guidelines and legal standards. These audits can identify risks and recommend corrective actions, providing a valuable service in the quest for ethical AI. Similarly, certification programs can offer a seal of approval for AI systems that meet high ethical and technical standards, reassuring stakeholders of their reliability and trustworthiness.

The financial implications of accountability cannot be overlooked. Building and maintaining accountable AI systems involves investments in resources, time, and expertise. Businesses must weigh the costs of implementing ethical practices against the potential risks of not doing so, which can include legal penalties, loss of consumer trust, and reputational damage. In this context, accountability is not merely a moral or legal obligation but a strategic business imperative.

International collaboration adds another layer of complexity and importance in defining accountability. AI systems do not recognize geographical boundaries, making international cooperation essential for establishing consistent accountability standards. Such efforts are already underway, with organizations like the United Nations and the European Union working on global AI ethics guidelines. These international standards aim to harmonize practices across borders, ensuring that accountability is maintained regardless of where an AI system is deployed.

In conclusion, defining accountability in AI systems is an intricate, multi-layered endeavor that touches upon technical, ethical, legal, user-centric, socio-technical, regulatory, financial, and international dimensions. Each of these aspects contributes to a comprehensive understanding of what it means to be accountable in the realm of artificial intelligence. As AI continues to evolve, so too must our approaches to accountability, adapting to new challenges and opportunities while steadfastly adhering to core ethical principles. Ultimately, a well-defined accountability framework is not just a safeguard against potential risks but a pathway to achieving the immense promise of AI for the betterment of humanity.

Legal and Ethical Standards

In the realm of AI systems, legal and ethical standards comprise an essential framework that guides developers, corporations, policymakers, and researchers in the responsible creation and deployment of these technologies. As AI continues to evolve and integrate deeper into various sectors, the need for robust legal and ethical mechanisms can't be overstated. Addressing these concerns ensures that AI systems contribute positively to society while minimizing potential harms. The acceleration of AI's capabilities calls for a synchronized effort to align legal regulations and ethical norms.

A significant aspect of legal and ethical standards in AI pertains to the concept of responsibility. Who is held accountable when an AI system makes a mistake? Should responsibility lay solely on the creators, or is there a shared burden between developers, users, and organizations? Legal frameworks aim to delineate these responsibilities clearly, but existing laws often lag behind the rapid pace of AI advancement. Thus, evolving these frameworks to include AI-specific scenarios is imperative.

Legal compliance is non-negotiable for organizations deploying AI systems. Regulations such as the General Data Protection Regulation (GDPR) in Europe have set high standards for data protection and privacy. These regulations are stepping stones toward comprehensive legal structures that govern AI, but gaps still exist. To fill these gaps, policymakers must engage in continuous dialogue with AI experts, ethicists, and the public to draft laws that address contemporary issues while being adaptable to future advancements.

Ethical considerations often stem from philosophical debates that have been ongoing for centuries. Concepts of fairness, justice, and human welfare are central to these discussions. In AI, these concepts translate into practical guidelines for the design, implementation, and monitoring of systems. For instance, ethical guidelines emphasize the importance of bias mitigation to ensure fairness. AI systems should not perpetuate existing social inequalities but rather aim to bridge these gaps.

A pivotal aspect of AI ethics involves the principle of beneficence, which insists that AI systems should do good and contribute positively to society. This principle raises questions about the deployment of AI in sensitive domains like healthcare, law enforcement, and education. Developers and organizations must navigate these sectors with a heightened sense of responsibility, ensuring that the benefits of AI are maximized without compromising ethical standards.

Transparency and accountability intersect significantly within legal and ethical frameworks. Laws and ethical guidelines often mandate that AI decision-making processes be transparent. This transparency allows for accountability as stakeholders can understand, scrutinize, and appeal decisions made by AI systems. Explainable AI models are a response to this demand, aiming to demystify how AI arrives at specific conclusions.

The ethical principle of autonomy underscores the importance of human agency in interactions with AI. AI systems should augment human decision-making rather than replace it entirely. Maintaining human oversight is crucial, especially in high-stakes environments like autonomous driving or medical diagnostics. Legal standards often enshrine this principle to safeguard human rights and prevent the over-reliance on automated systems.

Ethical standards also encompass issues of informed consent. When AI systems interact with individuals, it's vital that these individuals are aware and consent to how their data is being used. Ethical guidelines mandate clear communication regarding data collection, processing, and sharing practices. Transparency in these practices fosters trust and upholds the ethical treatment of data subjects.

The precautionary principle is another cornerstone of ethical AI. This principle suggests that in the face of uncertainty, caution should be the guiding force. When deploying AI technologies, especially in novel areas, developers and policymakers must evaluate potential risks and adopt a cautious approach. The aim is to anticipate and mitigate negative outcomes before they occur, thereby safeguarding the public and the environment.

Incorporating cultural sensitivity into legal and ethical standards is equally important. AI systems are deployed globally and often interact with a diverse user base. Ethical standards should respect cultural diversity and ensure that AI systems do not perpetuate cultural insensitivities or biases. Legal frameworks can support this through regulations that promote inclusive design practices and audit AI systems for cultural biases.

Ethical considerations in AI also extend to the environmental impact of these technologies. The computational power required for training advanced AI models has a significant carbon footprint. Ethical guidelines encourage the development of energy-efficient AI technologies and the adoption of sustainable practices. Legal standards may eventually enforce regulations to limit the environmental impact of AI, ensuring that technological advancement doesn't come at the expense of ecological well-being.

The intersection of AI with mental and emotional well-being is another area meriting legal and ethical scrutiny. AI systems, particularly those engaging in human-like interactions, must be designed to respect and enhance users' mental health. Ethical standards advocate for the humane treatment of individuals interacting with AI, and legal frameworks may establish protections against psychological manipulation or exploitation by AI systems.

Finally, the global nature of AI development and deployment necessitates international cooperation. Ethical standards and legal regulations need to transcend national borders and foster global collaborations. Organizations like the United Nations and the European Union are already taking steps to create international frameworks for AI governance. These collaborative efforts aim to harmonize regulations and ensure that AI development aligns with universal ethical principles.

Legal and ethical standards for AI are not just regulatory hurdles; they are foundational to the responsible advancement of this transformative technology. By embedding these standards into the very fabric of AI development, we can pave the way for systems that are not only innovative but also trustworthy, fair, and beneficial for all. The synthesis of rigorous legal compliance and robust ethical guidelines will create an environment where AI can reach its fullest potential while safeguarding human values and societal norms.

Case Studies

In exploring the landscape of accountability in AI systems, real-world case studies serve as invaluable milestones. They offer nuanced perspectives that theory alone cannot provide, showing both the pitfalls and potentials of AI technologies in various sectors. These stories act as both cautionary tales and sources of inspiration, outlining the path forward for developing responsible AI systems.

One of the most illustrative examples of the challenges and responsibilities associated with AI is the deployment of facial recognition technology by law enforcement agencies. In 2019, The Metropolitan Police Service in London decided to integrate real-time facial recognition in public surveillance cameras. The pilot program aimed to enhance public safety by identifying wanted criminals in real-time.

However, the project wasn't without its controversies. Several civil liberty organizations raised concerns about the accuracy and fairness of the technology. Reports surfaced that the system had alarmingly high rates of false positives, especially among minority populations. An evaluation from the University of Essex found that out of 42 matches generated by the system, only eight were correct. This revealed a significant flaw in the technology and underscored the ethical dilemma of deploying such systems without robust accountability measures. The public outcry and legal challenges that followed led to increased scrutiny and demands for transparency and public debate before further deployment.

The example of Amazon's AI recruiting tool further underscores the corporate world's struggle with accountability in AI. In an effort to streamline its hiring process, Amazon developed an AI system designed to review job applicants' resumes and identify top candidates. But by 2018, Amazon had to scrap the project. The AI had been trained on resumes submitted to the company over a ten-year period, most of which came from men. As a result, the AI developed a bias against female candidates. Despite attempts to reprogram the tool, it continued to exhibit discriminatory behavior. This highlighted the significant risk associated with inadequate training data and the need for continuous monitoring and auditing of AI systems to ensure they align with ethical standards.

In the healthcare sector, the case of IBM's Watson for Oncology has been both promising and problematic. Watson was developed to assist physicians in diagnosing cancer and recommending treatments. Initially, expectations were high; it was believed that Watson could revolutionize cancer care by providing highly personalized treatment options based on the latest medical research and patient data. However, implementation proved to be more challenging. Reports indicated that Watson frequently provided treatment recommendations that were incorrect or unsafe, likely due to incomplete or improperly curated data sources used for training. The gap between expectation and actual performance underscored the necessity for stringent validation processes and the importance of maintaining a feedback loop where clinicians can continuously evaluate and improve the AI's output.

Another revealing case comes from the financial sector. Zest-Finance, a tech company that builds machine learning tools for underwriting loans, attempted to create an AI-driven credit scoring model aimed at reducing bias inherent in traditional credit scoring systems. By leveraging various non-traditional data points, the goal was to provide fairer lending assessments. While initial results were promising, scrutiny revealed that the AI system could still perpetuate hidden biases, given that some alternative data points correlated with socio-economic backgrounds that had historically been marginalized. This case highlights the complexity of ensuring fairness in AI models and the importance of ongoing oversight and adjustments.

The realm of autonomous vehicles also provides pertinent case studies. Tesla's Autopilot has been at the forefront of public and regulatory attention. In one notable incident, a Tesla car operating on Autopilot crashed into a stationary vehicle, resulting in a fatality. The incident raised urgent questions about the responsibilities of manufacturers, the need for clear guidelines, and robust testing protocols to ensure public safety. It has led to more stringent regulatory requirements and increased transparency in reporting and evaluating the performance and safety records of autonomous systems.

Let's not overlook the role of AI in predictive policing, a domain rife with ethical complexities. The case of PredPol, a company that developed predictive policing software, highlighted the potential for both groundbreaking crime prevention and significant ethical missteps.

PredPol utilized historical crime data to predict where crimes were likely to occur, thus enabling police departments to allocate resources more effectively. However, studies later revealed that the system disproportionately targeted minority communities. The software amplified existing biases in policing practices and led to an uneven distribution of law enforcement efforts. This case emphasizes the critical need for accountability in the deployment of AI systems that can significantly impact human lives and societal structures.

In academia, the 2020 controversy involving Springer Nature and the publication of AI-generated papers exposed another dimension of accountability. The publisher had to retract multiple papers after it was discovered that they included nonsensical text generated by an AI writing tool. These instances exposed weaknesses in peer review processes and underscored the necessity for universities and publishers to implement stringent guidelines and verification protocols when dealing with AI-generated content.

Then there's the intriguing case of DeepMind's AlphaGo, which defeated top human Go players in 2016. While widely celebrated as a groundbreaking achievement in AI research, the event also sparked discussion about the accountability of AI systems in competitive environments. The complexity of AlphaGo's decision-making process, which even its creators found challenging to fully explain, raised significant questions about transparency and the need for explainable AI, especially in applications where understanding the decision process is crucial.

Looking at these diverse examples, it's evident that accountability in AI systems is not merely a theoretical concept but an urgent, multi-faceted challenge that spans industries and applications. Each case study demonstrates specific pitfalls and highlights the complexity of developing AI systems that are both innovative and ethically sound.

To move forward, stakeholders across various sectors must internalize the lessons learned from these cases. Creating a collaborative environment where developers, policymakers, and affected communities participate in the conversation is essential. By fostering an atmosphere of transparency, ongoing dialogue, and rigorous oversight, we can build AI systems that genuinely serve humanity, mitigate risks, and promote fairness and accountability.

As we navigate these uncharted waters, the real-world implications of AI's integration into society become increasingly apparent. The case studies discussed emphasize that accountability is not a mere add-on but a foundational element of responsible AI development. These narratives serve as stark reminders of what's at stake and the collective effort required to harness AI's potential responsibly.

CHAPTER 6:
AI IN THE WORKFORCE

The rise of AI in the workforce presents profound opportunities and challenges that demand our attention and thoughtful action. As automation accelerates, many jobs traditionally performed by humans are being transformed or even replaced by AI systems. This shift brings the promise of increased efficiency and innovation, yet it also raises pressing ethical questions about employment, economic disparity, and the future of work. It's crucial to consider how we can harness AI to complement human labor rather than displace it, ensuring that technological advancements lead to shared prosperity. Preparing for an AI-driven workforce requires a multi-faceted approach, including robust education and retraining programs, ethical standards for AI deployment, and policies that promote equitable economic growth. By addressing these challenges head-on, we can create a future where AI serves the greater good, empowering workers and enhancing societal well-being.

Automation and Employment

As artificial intelligence (AI) continues to grow in capability and reach, its impact on employment is becoming increasingly apparent. The integration of AI-driven automation in various industries raises significant questions about the future of work. While AI offers opportunities for increased efficiency and productivity, it also poses challenges that could reshape entire sectors and redefine the concept of employment itself.

Automation powered by AI is already disrupting traditional jobs. For instance, routine and repetitive tasks in manufacturing, data processing, and customer service can be efficiently handled by AI systems. These tasks, often monotonous and susceptible to human error, are prime candidates for automation. Businesses are eager to adopt AI solutions to cut costs and improve performance, leading to a significant shift in labor dynamics.

However, the rise of AI-driven automation doesn't mean an outright elimination of jobs. Instead, it often results in the transformation of existing roles. Employees may find themselves transitioning from task-oriented work to more complex, analytical responsibilities. For instance, factory workers might move from performing manual assembly to overseeing automated systems and ensuring their proper functioning.

A key concern is the pace at which these changes are happening. Rapid advancements in AI technology mean that automation can outstrip the ability of workers to adapt. Industries that once relied heavily on human labor face the urgent need to re-skill and up-skill their workforce. This requires not just individual effort but also concerted actions from governments, educational institutions, and businesses to provide the necessary training opportunities.

While lower-skilled positions are more vulnerable to automation, AI also brings about new job categories. Roles like machine learning engineers, data scientists, and AI ethicists are in demand. These positions require specialized knowledge and creativity, underscoring the need for a highly educated workforce. As such, the educational paradigm must shift to emphasize STEM (Science, Technology, Engineering, Mathematics) disciplines and foster continuous learning.

Some economists argue that AI could lead to job polarization. High-skilled jobs requiring advanced technical expertise will proliferate, while low-skilled jobs could diminish, leaving a gap in middle-skill employment. This polarization threatens to exacerbate income inequality if not managed properly. Policymakers need to address these socioeconomic imbalances to ensure fair distribution of AI's benefits.

Interestingly, automation doesn't just affect blue-collar jobs; it also impacts white-collar professions. Legal services, financial advising, and even journalism are areas where AI is making inroads. Automated systems can analyze legal documents, provide financial recommendations, and generate news articles with remarkable speed and accuracy. This shift redefines professional roles and necessitates a reevaluation of how we perceive skilled labor.

Despite the challenges, automation presents an opportunity to enrich human work experiences. By offloading monotonous tasks to AI, employees can focus on more creative, strategic, and interpersonal aspects of their jobs. This enhanced focus on human-centric tasks could improve job satisfaction and foster innovation. Creating a symbiotic relationship between AI and humans will be crucial moving forward.

The ethical implications of automation in employment cannot be ignored. Businesses implementing AI-driven automation must consider the human cost. Mass layoffs and unemployment spur social unrest and economic instability. Companies have a moral responsibility to provide support for displaced workers. Initiatives might include offering severance packages, career counseling, and financial support for continued education.

Ethical considerations also extend to the development and deployment of AI systems. Building AI that complements human work rather than replaces it can strike a balance between innovation and employment. Transparent communication about automation plans within organizations can help mitigate fears and resistance among employees.

Another vital aspect of automation's impact on employment is the role of regulation. Governments must enact policies that encourage fair labor practices in an AI-driven economy. This includes advocating for minimum wage adjustments, monitoring working conditions, and ensuring that workers have a voice in the automation process. Public-private partnerships could foster a collaborative approach to developing frameworks that balance technological progress with social welfare.

Moreover, the effects of AI automation are not uniform across the globe. Developing countries, which often rely heavily on low-skilled labor, might face significant economic disruptions. Conversely, countries with robust educational systems and flexible labor markets may adapt more readily. International cooperation and knowledge-sharing can help bridge these disparities and promote inclusive growth.

In conclusion, automation and employment represent a complex intersection of technology, economics, and ethics. While AI-driven automation has the potential to revolutionize industries and improve efficiency, it also presents challenges that require thoughtful consideration and proactive management. The future of work hinges on how society navigates these changes, balancing innovation with the well-being of the labor force. Only through collaborative efforts can we ensure that AI enriches rather than diminishes human work.

Ethical Considerations

The integration of artificial intelligence (AI) in the workforce raises several ethical considerations that demand our immediate and thoughtful attention. As AI progressively penetrates various industries, it becomes imperative to examine the broader implications of its deployment. Ethical challenges abound, ranging from the disruption of employment to concerns over fairness, transparency, and accountability.

In the context of automation, one major ethical consideration is the potential for job displacement. While AI has the capacity to enhance productivity and generate new job opportunities, it also threatens to render certain roles obsolete. This scenario raises significant ethical concerns about income inequality and the potential exacerbation of social disparities. The workforce must be prepared for this shift, and strategies such as reskilling and upskilling will be crucial in mitigating the adverse effects of automation.

Moreover, there's a pressing need to consider how decisions made by AI systems are perceived by the workforce. When AI algorithms dictate significant aspects of employment—ranging from hiring and performance evaluations to promotions and terminations—it is essential to ensure these decisions are fair and unbiased. Transparency in algorithmic decision-making becomes not just a technical challenge, but a moral imperative. Employees need to trust that AI-driven processes are both just and explainable.

Another ethical concern is the potential erosion of human agency and autonomy in the workplace. As AI systems become more prevalent, there is a risk that human workers may feel increasingly disempowered. Individuals who are constantly monitored and evaluated by AI may experience heightened stress and reduced job satisfaction. The design and deployment of AI in the workforce should aim to enhance, rather than undermine, human dignity and autonomy.

On a broader scale, the ethical deployment of AI in the workforce must consider the implications for society as a whole. The benefits of increased efficiency and productivity brought about by AI should not come at the expense of societal well-being. Policymakers, technologists, and business leaders must collaborate to create frameworks that ensure the equitable distribution of AI's advantages. This might involve developing robust social safety nets and policies that support workers impacted by AI-driven changes.

Data privacy is another pivotal ethical consideration. AI systems in the workplace often rely on large volumes of personal data to function effectively. This raises significant concerns about the protection and misuse of employee data. Employers must ensure that data collection practices are transparent and that rigorous safeguards are in place to protect sensitive information. Furthermore, employees should have the right to know how their data is being used and to give informed consent.

The potential for bias in AI systems is a critical ethical issue that directly impacts the workforce. If unaddressed, bias in algorithms can perpetuate and even exacerbate existing inequalities. For instance, hiring algorithms trained on biased historical data may favor certain demographics over others, leading to unfair hiring practices. Organizations must invest in auditing and refining their AI systems to identify and eliminate such biases. This involves ongoing vigilance and a commitment to fairness and inclusivity.

Ethical considerations must also extend to the broader societal impact of AI integration in the workforce. The transformation ushered in by AI has far-reaching consequences that go beyond individual workplaces. For instance, the displacement of jobs in one sector can have ripple effects on community stability and economic health. A holistic approach that considers these broader societal impacts is necessary to ensure a just and equitable transition.

Another facet of ethical consideration is the alignment of AI systems with human values. The development and deployment of AI in the workforce should be guided by ethical principles that prioritize human welfare, respect for individual rights, and the promotion of social good. Organizations should establish ethical guidelines and frameworks that govern the use of AI, ensuring that these systems serve to enhance human flourishing rather than undermine it.

Responsibility and accountability in the use of AI within the workforce are paramount. Organizations deploying AI must be accountable for the outcomes of their AI systems. This involves not only ensuring that AI systems are designed and implemented ethically but also addressing any negative consequences that arise. Accountability mechanisms might include transparent reporting, external audits, and the establishment of clear lines of responsibility.

Finally, the ethical deployment of AI in the workforce necessitates a commitment to continuous learning and adaptation. As technology evolves, so too must our ethical frameworks and practices. Organizations must be proactive in staying abreast of new developments and continually reassessing the ethical implications of their AI systems. This ongoing process of learning and adaptation will be crucial in navigating the complex ethical landscape of AI in the workforce.

In conclusion, the ethical considerations surrounding AI in the workforce are multifaceted and complex. Addressing these challenges requires a collaborative effort from all stakeholders, including developers, policymakers, business leaders, and workers. By fostering an ethical approach to AI deployment, we can harness the potential of AI to create a more just and equitable future for all.

Preparing for the AI-Driven Workforce

As artificial intelligence continues to embed itself into our daily lives, its impact on the workforce is both monumental and inevitable. The advent of AI disrupts traditional job roles, creates new opportunities, and initiates a shift that ushers in an era of collaboration between humans and machines. Preparing for this AI-driven workforce is not merely about technological adaptation but also about embracing ethical practices, re-skilling, and fostering a culture of continuous learning.

Firstly, it's crucial to identify the industries most likely to be transformed by AI. Sectors such as manufacturing, healthcare, finance, and customer service are poised for significant changes. AI can automate repetitive tasks, analyze vast datasets, and even assist in decision-making processes. However, this automation raises concerns about job displacement, exacerbating the need for a proactive approach to workforce development.

Reskilling and upskilling are pivotal strategies in this transformation. Companies and educational institutions must collaborate to create curricula that address the skills gap. Programs focusing on AI literacy, data analysis, coding, and cybersecurity are essential. Beyond technical skills, there is a growing emphasis on cultivating soft skills like critical thinking, problem-solving, and emotional intelligence. These human attributes become invaluable when paired with AI's capabilities, enabling workers to tackle complex, creative, and non-routine tasks that machines are not yet equipped to handle.

Governments and policymakers have a significant role to play in this transition. By implementing supportive regulations and incentivizing lifelong learning, they can foster an environment conducive to workforce adaptability. Public policies might include tax benefits for companies investing in employee training programs, grants for educational initiatives, and the promotion of public-private partnerships that drive innovation in workforce development.

Another critical aspect of preparing for an AI-driven workforce is ensuring ethical use and deployment of AI technologies. Developers and organizations must adhere to ethical guidelines that prioritize transparency, fairness, and accountability. This means creating AI systems that are explainable, unbiased, and aligned with societal values. Ethical AI development can build trust among employees, who may be skeptical or fearful of AI's role in their jobs.

Employee involvement in the AI integration process cannot be overstated. Inclusive dialogue between management and the workforce about the goals, processes, and implications of AI adoption ensures a smoother transition. Regular training sessions, workshops, and transparent communication build an informed workforce that feels empowered rather than threatened by AI.

Beyond organizational strategies, individuals must take responsibility for their career development. In an AI-driven world, taking initiative to learn and adapt becomes a personal mandate. Online courses, professional certifications, and participation in community learning groups can provide individuals with the knowledge and skills needed to stay relevant. This proactive approach to personal development fosters resilience in the face of technological changes.

Collaboration between humans and AI represents a new paradigm of workforce dynamics. AI should be viewed as a tool that augments human capabilities rather than replaces them. For example, in healthcare, AI algorithms can analyze medical images faster and more accurately than humans, but the final diagnosis and patient care decisions still require human expertise. This synergistic interaction can lead to more efficient workflows, improved decision-making, and enhanced productivity across various sectors.

Furthermore, organizations must prioritize data security and privacy in AI deployment. With the rise of AI, vast amounts of data are being generated and utilized, raising concerns about data breaches and misuse. Companies need to implement robust security measures and comply with data protection regulations to protect sensitive information. Equipping employees with knowledge about data ethics and cybersecurity practices is also crucial to safeguard against potential risks.

The **mental and emotional well-being** of employees during this transition is another important consideration. The fear of job displacement and the pressure to continuously adapt can contribute to stress and anxiety. Therefore, organizations should prioritize mental health resources and create a supportive work environment. Employee assistance programs, counseling services, and fostering a culture of open dialogue about workplace stress can help mitigate these challenges.

AI's influence on the workforce also raises questions about income inequality and social justice. As high-skilled jobs become more in demand and low-skilled jobs more automated, there's a risk of widening the economic divide. Policymakers must address these issues by ensuring equitable access to AI education and opportunities. Social safety nets, progressive taxation policies, and universal basic income (UBI) are potential solutions to consider in mitigating the socioeconomic impact.

In conclusion, preparing for the AI-driven workforce requires a multifaceted approach that includes reskilling, ethical AI development, individual responsibility, and supportive policies. By embracing these strategies, we can harness the benefits of AI while minimizing its risks. The future of work in an AI-driven world holds great promise, but realizing this potential necessitates proactive and thoughtful preparation.

CHAPTER 7:
AI IN HEALTHCARE

As we dive into the realm of healthcare, the integration of AI stands as both a beacon of transformative potential and a crucible of ethical challenges. The advantages are monumental: from revolutionizing diagnostics with unparalleled precision to creating personalized medicine tailored to individual genetic blueprints. Yet, with these advancements come profound ethical dilemmas. How do we balance innovation with patient privacy? The stakes are high, as biased algorithms can exacerbate healthcare disparities rather than alleviate them. Equally pivotal is the question of accountability when AI systems err in critical decisions. This chapter unravels the complexities of deploying AI in healthcare, detailing not only the opportunities that lie ahead but also the rigorous ethical and practical standards necessary to safeguard patient welfare and ensure equitable access to these groundbreaking technologies. By examining case studies and best practices, we will explore ways to harness the power of AI responsibly, ensuring it serves humanity's best interests in the most intimate and crucial corner of our lives—our health.

Benefits and Risks

The impact of artificial intelligence (AI) in healthcare is profound, presenting a myriad of benefits that can transform patient care, enhance operational efficiencies, and contribute to ground-breaking medical research. However, these advantages come with significant risks that warrant careful consideration by developers, policymakers, and healthcare providers alike.

One of the most compelling benefits of AI in healthcare is its potential to revolutionize diagnostics. AI algorithms, fed with colossal datasets, can identify patterns and anomalies in medical imaging with remarkable accuracy. This capability means faster and more precise diagnoses, helping to detect diseases like cancer at earlier stages when they are more treatable. The implications for patient survival rates and quality of life improvements are substantial.

Moreover, AI can personalize treatment plans in unprecedented ways. By analyzing individual patient data, including genetic information and health history, AI systems can recommend tailored treatment protocols. This personalized approach extends to predicting patient responses to specific drugs, thereby reducing the trial-and-error period and associated side effects. Such precision medicine holds promise for chronic disease management, significantly enhancing patient outcomes.

Operational efficiencies represent another significant advantage of AI in healthcare. Automation of administrative tasks, such as scheduling, billing, and patient registration, can save valuable time and reduce human errors. In a healthcare setting where resource optimization is crucial, AI-driven systems can streamline workflows, allowing medical staff to focus more on direct patient care. This shift could lead to better doctor-patient interactions and improved healthcare delivery.

A further strength of AI is in medical research. Powerful AI tools can analyze vast amounts of scientific literature, clinical data, and experimental results much faster than traditional methods. These tools can identify promising research pathways, aiding in the discovery of new drugs and therapies. The acceleration of research timelines through AI can lead to quicker development of effective treatments, ones that could combat diseases that currently have limited options.

Despite these tremendous benefits, the incorporation of AI in healthcare is fraught with risks that must be navigated carefully. One of the primary concerns is data privacy. Healthcare data is extremely sensitive, and the integration of AI systems requires the collection and processing of vast amounts of personal health information. Without stringent data protection measures, this data could be vulnerable to breaches, leading to potential misuse and significant harm to patient privacy.

Another critical risk is the potential for bias in AI algorithms. Since AI systems learn from historical data, they can perpetuate existing biases present in that data. In a healthcare context, biased algorithms could result in unequal treatment recommendations that disproportionately affect certain demographic groups. For example, if an AI system is trained predominantly on data from a specific population, it may not perform as well for patients from different backgrounds, leading to disparities in healthcare quality and outcomes.

There's also the issue of AI explainability, or the lack thereof. Many advanced AI algorithms, especially deep learning models, function as "black boxes," making it difficult to understand how they make decisions. In healthcare, this opacity can be particularly problematic. Doctors and patients may be reluctant to trust or follow AI-generated recommendations if they don't understand the rationale behind them. This lack of transparency can hinder the adoption of AI technologies and undermine their potential benefits.

Furthermore, the reliability and robustness of AI systems in healthcare are paramount, as errors or malfunctions can have life-or-death consequences. AI systems must undergo rigorous testing and validation to ensure they perform reliably under diverse conditions. The failure to adequately test these systems could lead to incorrect diagnoses, inappropriate treatments, or other serious medical errors, eroding trust in technological advancements.

Ethical considerations are equally vital and complex. The delegation of certain medical decisions to AI raises questions about responsibility and accountability. Who is responsible if an AI system makes a mistake? The developers, the healthcare providers, or the institution employing the technology? Clear guidelines and robust ethical frameworks are needed to address these questions and to ensure that the usage of AI in healthcare aligns with broader ethical standards.

As we look at the integration of AI in healthcare, it is evident that while the promised advancements are significant, the associated risks are equally substantial. Thus, a balanced approach is necessary—one that embraces innovation while meticulously addressing the ethical, privacy, and operational challenges. Collaboration among technologists, healthcare professionals, ethicists, and policymakers will be crucial in developing and deploying AI systems that enhance healthcare without compromising safety, fairness, or privacy.

Finally, public perception and trust in AI technologies play a crucial role in their acceptance. Patients must be educated about the benefits and potentials of AI while being reassured that their data and privacy are being protected. Transparent communication and involving patients in discussions about AI applications in their care can help build this trust. Developing AI-literate healthcare professionals is also essential, as they will be at the forefront of interfacing between technology and patient care.

In conclusion, AI in healthcare offers transformative possibilities, but its deployment must be undertaken with caution and a thorough understanding of associated risks. By addressing the benefits and risks comprehensively, we can navigate the frontier of AI in healthcare responsibly and ethically, ensuring that this powerful technology serves to improve human well-being without compromising the ethical foundations and trust upon which the healthcare profession is built.

Ethical Dilemmas

In the rapidly evolving arena of healthcare, artificial intelligence (AI) offers groundbreaking possibilities. From early diagnostics to personalized treatment plans, AI systems promise to revolutionize the way we understand and practice medicine. However, along with these transformative benefits come significant ethical dilemmas that must be addressed to avoid potential harms.

One fundamental ethical dilemma revolves around the balance between technological advancement and patient privacy. AI technologies often require vast amounts of data to function effectively. This data includes sensitive patient information, raising pertinent questions about consent and confidentiality. For instance, do patients fully understand how their data will be used, and do they have the option to opt out?

Another point of contention lies in the bias embedded within AI algorithms. Bias in AI can manifest in various forms—racial, gender, socioeconomic, and more. In healthcare, biased algorithms could lead to misdiagnoses or unequal treatment among different demographic groups. It's crucial to scrutinize the data used to train these algorithms and to implement checks that identify and mitigate these biases.

Transparency is also a pressing ethical issue. When an AI system makes a decision about patient care, can it explain how it arrived at that decision? This lack of "explainability" can make it difficult for healthcare providers to trust and rely on AI, thus hindering its integration into healthcare systems. Moreover, it raises alarming concerns about patient autonomy and informed consent.

The potential for AI to make life-and-death decisions introduces a level of urgency to these ethical concerns. For example, who is accountable if an AI system makes an incorrect diagnosis or suggests an ineffective treatment? The lines of responsibility become blurred, challenging the conventional understanding of medical accountability. This can complicate legal and ethical repercussions, leaving both healthcare providers and patients in precarious situations.

Further, the deployment of AI in healthcare settings is not just a technical issue; it's a social one. Raising questions about the societal implications can help us understand the broader impact of AI on healthcare access and equity. For example, will these advanced technologies be available only to those who can afford them, thereby widening existing healthcare disparities? It's imperative to ensure that AI benefits are equitably distributed across different socio-economic and geographic populations.

Another contentious issue is the potential dehumanization of patient care. While AI can efficiently process data and suggest treatment plans, it lacks the empathy and human touch that is critical in healthcare. Can AI ever replace the nuanced understanding and emotional support offered by human healthcare providers? This concern highlights the need for a balanced approach, integrating AI's computational abilities with human empathy and judgment.

Deploying AI in healthcare also challenges existing ethical frameworks and necessitates new guidelines tailored to these emerging technologies. Policymakers and healthcare providers need to collaborate on setting these standards, ensuring they are adaptable to the fast pace of technological innovation. International cooperation can also play a crucial role in establishing uniform ethical guidelines, ensuring that AI's promise in healthcare does not get overshadowed by preventable pitfalls.

The complexity of these dilemmas underscores the importance of a multi-disciplinary approach in tackling ethical issues in AI healthcare. This requires input not only from technologists and healthcare providers but also from ethicists, sociologists, and even patients. Engaging a broad spectrum of voices can lead to more holistic and inclusive solutions, mitigating the risks while amplifying the benefits.

An exemplar of tackling these ethical dilemmas can be seen in community-based participatory research (CBPR) initiatives, where researchers work directly with communities to develop AI solutions tailored to their specific needs and ethical concerns. Such initiatives promise a more inclusive approach to developing AI technologies, ensuring they are aligned with the values and necessities of diverse populations.

In conclusion, while AI holds transformative potential for the healthcare sector, it is accompanied by ethical dilemmas that demand meticulous attention. Striking the balance between innovation and ethical responsibility requires concerted effort, stringent standards, and multi-disciplinary collaboration. By addressing these ethical dilemmas thoughtfully, we can pave the way for responsible AI in healthcare, ensuring it serves as a boon rather than a bane.

Case Studies and Best Practices

Delving into the realm of AI in healthcare provides us with a treasure trove of case studies that highlight both the immense potential and the intricate ethical quandaries posed by these emerging technologies. Throughout these examples, the importance of effective and responsible AI deployment becomes clear. Attention to ethical considerations in patient data management, algorithmic fairness, transparency, and accountability can offer transformative benefits while safeguarding the trust and well-being of patients.

Consider the case of IBM Watson's foray into oncology. IBM Watson was designed to assist oncologists by analyzing vast amounts of medical literature to recommend cancer treatments. The system had access to an enormous pool of data, including medical journals, clinical trial outcomes, and patient records. Initially, the expectations were high. IBM Watson promised to revolutionize cancer care by providing personalized treatment recommendations. However, the outcomes were mixed. Some healthcare institutions reported positive outcomes, with Watson's recommendations aligning with expert opinions and sometimes even offering novel insights.

Yet, the implementation revealed significant caveats. For instance, a study at MD Anderson Cancer Center brought some of these limitations to light. The AI system struggled to navigate complex patient data that deviated from its training datasets. In certain scenarios, its recommendations were not always practical or effective. These findings underscored the importance of quality and representativeness of training data. The case study of Watson in oncology emphasizes the importance of ongoing evaluation and refinement of AI systems in healthcare. For AI to be a reliable aid in medical decision-making, continuous learning and tweaking of algorithms are essential, mandates requiring robust clinical validation processes.

In another impactful example, Google's DeepMind collaborated with the U.K.'s National Health Service (NHS) to leverage AI for the early detection of acute kidney injury through its Streams app. This partnership aimed to harness DeepMind's machine learning capabilities to analyze patient data, predict deterioration, and send alerts to clinicians for timely intervention. The application demonstrated substantial promise, offering the potential to save lives by addressing acute medical conditions before they escalate.

However, the initial roll-out faced severe critiques over patient data privacy issues. The Royal Free NHS Foundation Trust, involved in the partnership, was accused of failing to secure the necessary patient consent for data use, igniting a public debate over data privacy and ethical responsibility in AI applications. This incident served as a critical lesson in ensuring stringent compliance with data protection regulations and obtaining informed consent. Additionally, it revealed that the ethical use of AI in healthcare isn't just about technical efficacy but also about respecting patient autonomy and privacy.

Reflecting on best practices, institutions employing AI should prioritize a patient-centered approach, inclusive of transparency and stakeholder engagement. The integration of AI tools should always involve healthcare professionals in the loop. AI should augment, not replace, human expertise. This necessity is exemplified by the collaboration between radiologists and AI algorithms in interpreting mammograms. Several studies, including those conducted by MIT and Harvard, have revealed that a synergistic approach — where radiologists utilize AI tools for double-checking results — significantly improves accuracy rates in breast cancer detection. This collaboration underscores the value of combining machine precision with human intuition and expertise.

Moreover, fostering a culture of ethical responsibility within development teams is fundamental. Ethical AI development frameworks, such as those outlined by the Partnership on AI, advocate for diversity in training datasets to avoid perpetuating biases. One potent example comes from the development of AI-driven diagnostic tools for skin diseases. Initial versions of these algorithms were far more accurate in diagnosing conditions in lighter-skinned individuals due to disproportionate representation in the training data. Readdressing this bias involved incorporating a diverse array of skin types into the training sets, thus ensuring fairer and more accurate diagnostic capabilities across all demographic groups. This case highlights the importance of diversity and inclusiveness in data collection and algorithm training processes.

Another vital component is the thorough documentation and communication of AI decisions. Explainable AI (XAI) plays a crucial role here. Consider the use of AI in predictive analytics for patient outcomes in intensive care units. Systems developed by institutions like the Mayo Clinic utilize machine learning to forecast patient deterioration, enabling preemptive medical interventions. However, the efficacy and acceptance of such systems are heavily reliant on their transparency. Clinicians must understand not only the predictions but also the reasoning behind them. Employing explainable AI models ensures that healthcare providers can grasp the underlying logic, building trust and facilitating more informed decision-making processes. Detailed documentation also aids in auditing and improving these systems, paving the way for continuous enhancement and ethical compliance.

Regulatory frameworks and standard guidelines are indispensable in navigating the ethical landscape of AI in healthcare. The Food and Drug Administration (FDA) in the United States has been at the forefront of establishing protocols for the approval and monitoring of AI-driven medical devices. For instance, the FDA's regulation of AI-based diagnostic tools emphasizes the necessity of demonstrating clinical validity and ensuring real-world performance through post-market surveillance. This structured oversight ensures that AI applications meet stringent safety and efficacy standards before they can impact patient care. Moreover, ongoing assessments help address any emerging risks or unintended consequences.

Lastly, fostering interdisciplinary collaborations is paramount. The ethical deployment of AI in healthcare benefits profoundly from the insights and expertise of a diverse array of stakeholders. Including ethicists, legal experts, healthcare practitioners, data scientists, and patients in the development and implementation phases ensures a well-rounded approach to ethical considerations. These collaborative efforts help preempt potential ethical pitfalls and enhance the societal acceptance of AI technologies.

To conclude, the transformative potential of AI in healthcare is accompanied by significant ethical responsibilities. Through the careful examination of case studies and adherence to best practices, we can establish a balanced approach that maximizes benefits while mitigating risks. By committing to transparency, inclusivity, continuous evaluation, and interdisciplinary collaboration, we set the stage for responsible and ethical AI systems that can profoundly improve patient outcomes and healthcare services.

CHAPTER 8:
SECURITY AND ROBUSTNESS

Ensuring the security and robustness of AI systems is paramount in a world increasingly reliant on these technologies. Without these foundational attributes, AI, regardless of its capabilities, remains vulnerable to threats, ranging from cyber-attacks to subtle adversarial manipulations. The path to robust AI involves rigorous testing, constant vigilance, and the adoption of comprehensive security practices designed to anticipate and mitigate potential risks. Ethical AI development isn't just about creating advanced algorithms; it's about building resilient systems that can withstand malicious attempts to compromise their integrity. Developers, policymakers, and researchers must collaborate, adopting best practices that foster trustworthy AI. This collaborative effort will help ensure AI systems not only perform as intended but also enhance the safety and reliability upon which society increasingly depends.

Threats to AI Systems

In the rapidly advancing arena of artificial intelligence (AI), security and robustness are paramount concerns. The very ubiquity and integration of AI into diverse sectors—from healthcare to law enforcement—underscore its susceptibility to an array of threats. Understanding these threats is a cornerstone for developing robust and secure AI systems that can effectively withstand adversities and fulfill ethical commitments.

One of the primary threats to AI systems is the risk of adversarial attacks. These occur when malicious actors deliberately input deceptive data to manipulate or exploit an AI system's vulnerabilities. A seemingly innocuous image, subtly modified in ways imperceptible to human observers, can deceive an AI image recognition system into misclassifying it. This type of attack exposes a disconcerting flaw: AI systems often lack the nuanced contextual understanding that humans possess.

Then there's the issue of data poisoning. AI systems rely heavily on large datasets to learn and make predictions. If an attacker can introduce erroneous or malicious data during the training phase, the entire model's integrity is compromised. This undermines trust in AI outputs and can have wide-reaching implications—from financial fraud in automated trading systems to misdiagnosis in AI-driven healthcare modalities.

Another notable threat is that of model inversion and data theft. AI models often deal with sensitive information. Sophisticated techniques can potentially reverse-engineer a trained model to extract anecdotal data points, thereby compromising user privacy. This becomes particularly troubling when personal or proprietary information is at stake. Mechanisms deployed to protect data must evolve in tandem with these emerging threats to ensure continuous robustness.

The landscape of cybersecurity is not static. As AI technology progresses, so do the techniques employed by cybercriminals. One alarming trend is the use of AI itself to perpetrate attacks. Malicious actors could develop AI-driven malware that autonomously learns and adapts, making detection and mitigation increasingly difficult. This self-perpetuating cycle of AI-driven attacks leads to an escalating arms race between security experts and attackers.

Furthermore, AI systems are not immune to social engineering tactics. Social engineering exploits human psychology to manipulate individuals into divulging confidential information or performing actions that compromise security. As AI becomes more integrated into customer service, chatbots, and personal assistants, the risk of these systems being tricked into divulging sensitive information increases. The blend of human factors and technological vulnerabilities complicates the security landscape exponentially.

Supply chain attacks represent another vector of vulnerability. AI systems often depend on a complex supply chain of software components, data sources, and hardware devices. Compromising any element within this chain can introduce vulnerabilities into the final AI product. Ensuring the integrity of the supply chain and routinely auditing each component are critical steps in mitigating these risks.

Additionally, the threat posed by algorithmic bias and fairness affects the security and robustness of AI systems. If an AI system exhibits bias, whether through biased training data or flawed algorithmic design, it can produce inequitable outcomes that erode trust and invite scrutiny. This bias isn't merely an ethical concern but also a security risk. Malicious actors might exploit biased models to induce systemic failures or amplify social divides, causing broader destabilization.

There's also the matter of AI impersonation. Deepfakes and synthetic media create convincing but falsified versions of real individuals. These technologies can be weaponized to distort public discourse or impersonate individuals for nefarious purposes. The sophistication of AI-driven deepfakes makes them a potent threat, capable of undermining personal reputations, corporate credibility, and even national security.

Physical threats to AI systems should not be overlooked either. Autonomous systems, such as drones or self-driving cars, are particularly susceptible. Ensuring the physical security of such systems is as crucial as protecting their digital underpinnings. Any compromise, whether through direct tampering or environmental manipulation, can have dire consequences, including accidents or loss of life.

Finally, regulatory and compliance risks represent a unique category of threats. As global and local regulations evolve, AI systems must continuously adapt to remain compliant. Regulatory non-compliance can result in significant penalties and damage to reputation. Moreover, regulatory complexity can introduce implementation challenges that may, paradoxically, introduce new vulnerabilities.

In conclusion, the threats to AI systems are multifaceted and evolving. From adversarial attacks and data poisoning to social engineering and physical tampering, each threat vector serves as a reminder of the critical importance of security and robustness. Ensuring that AI systems are robust, secure, and ethically aligned demands vigilance, continuous adaptation, and a proactive approach to emerging threats. In addressing these challenges, we take a vital step towards realizing the full potential of AI in a manner that is secure, robust, and ethically responsible.

Ensuring Robustness

Ensuring the robustness of artificial intelligence systems isn't merely a technical challenge; it is a moral imperative. As we increasingly depend on AI for critical aspects of our lives, from healthcare to law enforcement, a system's reliability can directly impact human well-being. Robustness, in this context, refers to an AI system's ability to perform accurately and consistently under diverse and potentially unexpected conditions. The journey to achieving this robustness is multifaceted, involving rigorous testing, continuous monitoring, and ethical foresight.

Central to ensuring robustness is the practice of rigorous testing under a plethora of scenarios. AI systems must be exposed to a vast array of inputs during their training phase, ensuring they can handle edge cases and anomalous data. This extensive testing helps identify potential vulnerabilities and shortcomings that could be exploited or could fail under unique pressures. Such thorough vetting is akin to preparing a ship for all types of weather conditions before setting sail. Analogously, a robust AI system must be built to withstand and adapt to various operational "weather conditions." The goal is to emerge unscathed from not just expected but also unforeseen challenges.

Security and robustness are intrinsically connected. A robust AI system is inherently more secure because it resists manipulation and performs reliably even when exposed to adversarial conditions. This includes defending against attacks aimed at exploiting known or unknown vulnerabilities in the AI's algorithm. Here, adversarial training becomes a critical component. By intentionally exposing the AI to adversarial examples during the training phase, developers can reinforce the system's defenses, much like a vaccination prepares the human immune system to ward off diseases.

Monitoring is another cornerstone of robustness. Even after extensive testing, an AI system must be continuously observed to ensure it remains fault-resistant over time. Considering the dynamic environments in which AI systems operate, real-time monitoring helps in detecting and rectifying new vulnerabilities. Modern AI systems often come equipped with self-check protocols that alert administrators to discrepancies, ensuring that any deviation from expected behavior can be promptly addressed. This real-time oversight essentially acts like a watchful guardian, ever-vigilant and ready to intervene when necessary.

In building robust systems, feedback loops are indispensable. Post-deployment feedback allows developers to gain insights from real-world operations, offering opportunities for continuous improvement. When an AI system encounters a situation that it handles poorly, that feedback should be cycled back into the development process.

Such iterative enhancements help in tuning the AI for optimal performance, embodying the principle of learning from one's mistakes.

One practical approach to maintaining robustness is the implementation of fail-safes and redundancy. Just as a skilled pilot relies on multiple instruments and backup systems to navigate safely, an AI system should have redundant pathways to ensure it can operate even if one component fails. These redundancies can be computational (multiple algorithms checking each other) or systemic (alternative systems ready to take over in case of a failure). Implementing such fail-safes ensures that a single point of failure does not compromise the entire system's integrity.

Ethical foresight is paramount. Anticipating the potential ethical dilemmas and societal impacts of an AI system enhances its robustness. Ethical considerations must be interwoven with technical standards from the onset, ensuring that the AI system aligns not just with operational expectations but also moral principles. This includes evaluating how the system's decisions might affect different demographic groups and ensuring fairness and non-discriminatory practices are embedded in its core functions.

It's also vital to ensure that AI systems maintain transparency and explainability. A system that can clearly articulate its decision-making process is more amenable to scrutiny and thus more robust. When stakeholders, including end-users and regulators, can understand the rationale behind AI decisions, it becomes easier to trust and verify these systems. Transparent systems help in pinpointing faults quickly, facilitating rapid response and remediation, which are crucial for maintaining ongoing robustness.

Consider the challenges posed by changing data landscapes. As the world evolves, so too does the data that AI systems must process. An initially robust system can degrade over time if it doesn't adapt to new data trends or shifts in user behavior. Therefore, dynamic re-training and adaptive algorithms are pivotal. An AI system should not be viewed as a static product but as a living entity that grows and evolves with its environment. Regular updates and re-training protocols are essential for an AI system to maintain its robustness in real-world applications.

Collaboration among various stakeholders adds another layer of robustness to AI systems. By involving experts from different fields—developers, ethicists, legal advisors, and user representatives—one can ensure that multiple perspectives are considered during the development phase. This interdisciplinary approach can preemptively identify potential shortcomings and enhance the system's resilience. The inclusion of diverse viewpoints helps in creating a balanced and robust AI that is well-prepared for a broad range of scenarios and challenges.

In conclusion, ensuring robustness in AI systems is a complex but necessary endeavor. It requires a blend of rigorous testing, continuous monitoring, ethical foresight, and stakeholder collaboration. By adopting a comprehensive approach that integrates technical excellence with moral accountability, we can build AI systems that not only perform reliably but also ethically and responsibly. Such robustness is not just a technical achievement; it is a reflection of our commitment to creating AI systems that serve humanity with the highest standards of integrity and reliability.

Security Best Practices

Ensuring the security and robustness of AI systems is paramount in today's interconnected world. The surge in AI applications has not only transformed industries but has also introduced profound security implications. Security best practices are a cornerstone of maintaining trust, integrity, and functionality in AI systems, addressing vulnerabilities that might be exploited by malicious actors.

One of the foundational pillars of security best practices is the principle of "defense in depth." This involves creating multiple layers of defense that safeguard an AI system from various threats. By integrating multi-factor authentication, encryption, and regular security audits, organizations can create a robust defense mechanism. This layered security model ensures that if one defense mechanism fails, others will still protect the system, mitigating the risk of catastrophic failures.

Regular security audits and vulnerability assessments are not just a one-time activity but an ongoing process. These assessments help in identifying potential weak points within the system. By conducting routine penetration testing, organizations can simulate cyber-attacks and evaluate the robustness of their AI infrastructures. This proactive approach allows for the timely identification and patching of vulnerabilities before they can be exploited by attackers.

Data privacy and security are intertwined with the broader framework of ethical AI. Implementing robust encryption methods ensures that sensitive data remains protected during both storage and transmission. Encryption safeguards data from unauthorized access and breaches, reinforcing trust between AI systems and their users. Additionally, secure data storage solutions must comply with regulations, such as GDPR, to uphold privacy standards.

Another critical aspect of security best practices in AI is securing the software development lifecycle (SDLC). Embedding security measures right from the initial design phase ensures that robustness is ingrained within the system. Techniques like secure coding practices, code review, and static code analysis play a vital role in detecting and mitigating security vulnerabilities early in the development cycle. Moreover, utilizing frameworks and libraries that are regularly updated and maintained reduces the risk of introducing vulnerabilities into the system.

Access control mechanisms form the backbone of any secure AI system. Implementing role-based access control (RBAC) ensures that only authorized users have access to specific functions and data within the system. This principle of least privilege restricts users' access rights to the minimum necessary to perform their tasks, thereby minimizing the risk of insider threats. Additionally, maintaining an access control policy and regularly reviewing users' permissions helps in keeping the system secure.

Data integrity is another cornerstone of security best practices. Ensuring that the data used by the AI system is accurate and unaltered is crucial for maintaining trust and reliability. Techniques like hashing and checksums can be employed to verify the integrity of data, ensuring that it has not been tampered with. Furthermore, maintaining a secure backup strategy can help in recovering from accidental data loss or malicious data breaches.

AI systems must also be designed to detect and respond to anomalies in real-time. Intrusion detection systems (IDS) and intrusion prevention systems (IPS) can monitor network traffic and system behavior for any suspicious activity. These systems can trigger alerts or take automated actions to prevent potential threats, thereby enhancing the resilience of AI systems. Additionally, employing machine learning models for threat detection can add an extra layer of security by identifying patterns that traditional rule-based systems might miss.

Training personnel in security best practices is as important as the technical measures themselves. Employees at all levels must be aware of security protocols and the importance of maintaining them. Regular training sessions and awareness programs can help in creating a culture of security within the organization. By understanding the potential risks and the importance of following security protocols, employees can act as the first line of defense against potential threats.

Third-party risk management is an integral component of the security framework. Organizations often rely on third-party vendors for various services, thus expanding their risk landscape. Establishing stringent security requirements for these vendors and regularly auditing their compliance can significantly mitigate third-party risks. Contracts with clear security and privacy obligations help ensure that vendors adhere to the same security standards expected within the organization.

Comprehensive incident response plans are crucial for dealing with security breaches promptly and effectively. These plans should outline the steps to identify, contain, eradicate, and recover from security incidents. Regular drills and simulations can help in testing the effectiveness of these plans and preparing the organization's response team for real-world situations. Having a well-defined communication plan ensures that stakeholders are informed timely and accurately, maintaining transparency and trust.

Implementing a robust logging and monitoring system is vital for maintaining security and accountability. Logs provide a detailed account of system activities and can be invaluable during forensic investigations. Regularly reviewing these logs helps in identifying patterns or anomalies that might indicate security issues. Automated monitoring solutions can enhance this process by providing real-time alerts and actionable insights.

Collaborating with the wider security community can also bring substantial benefits. Engaging in information sharing with other organizations can help in staying abreast of the latest threats and defense strategies. Participating in security consortia and working groups fosters a collective approach to tackling global security challenges and contributes to a more secure AI ecosystem.

Finally, fostering a culture of continuous improvement ensures that security measures evolve with emerging threats. This involves staying updated with the latest advancements in cybersecurity, participating in industry forums, and continuously refining security practices based on lessons learned from past experiences. By adopting an attitude of vigilance and adaptability, organizations can better prepare for the dynamic nature of security threats in AI.

In conclusion, security best practices for AI systems encompass a comprehensive and multi-layered approach. From technical measures like encryption and access controls to organizational practices like training and incident response planning, each aspect plays a vital role in safeguarding AI systems against potential threats. By implementing these best practices, organizations can create secure, robust, and trustworthy AI systems that uphold ethical standards and foster user confidence.

CHAPTER 9:
AI IN LAW ENFORCEMENT

As we delve into the realm of law enforcement, the integration of artificial intelligence unveils both promise and peril. AI technologies can streamline investigative tasks, predict criminal activity, and enhance public safety. Yet, these advancements also raise profound ethical concerns. The use of AI for surveillance must grapple with issues of privacy intrusion and the potential for abuse of power. Furthermore, the algorithms driving these systems need to be transparent to avoid biases that could unfairly target certain populations. Balancing the undeniable benefits with these substantial risks requires a meticulous framework of guidelines to ensure responsible and fair use. Achieving this balance is not merely a technical challenge but a societal one, demanding concerted efforts from developers, policymakers, and communities to create an equitable future for AI-driven law enforcement.

Surveillance Issues

In an era when technology evolves at breakneck speed, the integration of Artificial Intelligence (AI) into law enforcement practices ushers in both unprecedented capabilities and complex ethical dilemmas. Chief among these challenges are the surveillance issues that arise when AI is employed to monitor, predict, and even prevent criminal activities. Surveillance, inherently intrusive, becomes even more contentious when enhanced by AI's capacity for extensive data analysis and interpretation.

Imagine a network of cameras throughout a city, each equipped with facial recognition capabilities powered by sophisticated AI algorithms. While such a system could significantly improve public safety by identifying criminals and missing persons in real-time, it also raises profound questions about privacy. Are citizens willingly sacrificing their rights to privacy for the sake of security? And if so, where does one draw the line?

AI-enabled surveillance also has the potential to exacerbate existing biases and inequalities. Studies have shown that facial recognition algorithms often exhibit biases, particularly against women and people of color. When these biased systems are deployed by law enforcement, they can lead to wrongful identifications and unfair targeting, thereby undermining public trust and perpetuating systemic injustices.

The scope of surveillance has expanded beyond traditional methods, thanks to AI. Law enforcement agencies are now able to analyze social media, track online behaviors, and monitor communications on an unprecedented scale. This vast amount of data collection often bypasses conventional legal protocols and oversight, leading to concerns about overreach and abuse.

A critical issue lies in the accountability of the AI systems used for surveillance. Who is responsible when an AI system makes an incorrect or biased assessment? Unlike traditional policing methods, AI decisions can be opaque, making it challenging to hold individuals or organizations accountable. This opacity leads to a lack of trust among the public and creates a vacuum in which ethical missteps are harder to identify and rectify.

The psychological impact of constant surveillance is another consideration. Knowing that one's actions are perpetually monitored can lead individuals to self-censor and inhibit free expression. This phenomenon, known as the "chilling effect," poses significant threats to democratic values, such as freedom of speech and assembly.

Moreover, the data gathered through AI surveillance systems poses security risks. The aggregation of sensitive information can become a target for cyberattacks, jeopardizing not only individual privacy but also national security. Ensuring the robustness and security of these systems is paramount, yet it remains a daunting challenge.

Policymakers and developers must work in tandem to create regulations that safeguard privacy without stifling technological innovation. Clear guidelines and strict oversight mechanisms can help strike this delicate balance, ensuring that AI-driven surveillance is conducted ethically and transparently.

Public awareness and education are crucial in this discourse. Citizens must be made aware of their rights and the ways in which their data is collected and used. Public consultations and transparent reporting can foster a more informed and engaged populace, which is essential for democratic accountability.

Finally, international cooperation is vital. Surveillance issues are not confined to national boundaries; they have global implications. By fostering international dialogue and sharing best practices, nations can collaboratively address the ethical, legal, and technical challenges posed by AI in surveillance, ensuring a more equitable and secure future for all.

In sum, while AI holds the promise of transforming law enforcement for the better, the surveillance issues it brings to the forefront demand our careful attention and action. By addressing these challenges head-on, we can harness the power of AI responsibly, safeguarding both our security and our fundamental rights.

Ethical Implications

The integration of artificial intelligence (AI) into law enforcement brings forth a myriad of ethical considerations that require vigilant attention. While the potential for AI to enhance public safety and streamline investigative processes is significant, the implications of its deployment call for a robust examination of ethical principles. The fundamental challenge lies in balancing the benefits of AI with potential risks to civil liberties. This balancing act necessitates a thoughtful approach to ensure that the deployment of AI technologies in law enforcement adheres to the highest ethical standards.

One of the paramount ethical concerns involves the potential for AI systems to perpetuate or even exacerbate existing biases. Law enforcement databases often contain biased data, the result of historical and systemic prejudices. When AI systems are trained on this data, they can inherit and amplify these biases, leading to unfair targeting of minority communities. The troubling consequences of such biased decision-making include wrongful arrests, increased surveillance, and erosion of trust between law enforcement and the communities they serve. It is imperative, therefore, to implement rigorous bias detection and mitigation strategies in AI systems used by law enforcement to ensure justice and fairness.

Transparency is another crucial factor in the ethical deployment of AI in law enforcement. Citizens have the right to understand how AI is influencing policing practices and decisions. A lack of transparency can lead to a perception of secrecy and undermine public trust. It is essential for law enforcement agencies to be open about the AI technologies they deploy, the data they use, and the policies governing their use. Clear communication about the functioning, limitations, and ethical safeguards of AI systems can help demystify the technology and foster public confidence.

The issue of accountability cannot be overlooked when discussing the ethical implications of AI in law enforcement. Who is held responsible when an AI system makes an erroneous or harmful decision? The opacity of many AI algorithms often complicates the assignment of accountability. Establishing clear lines of accountability ensures that human oversight is maintained and that there are mechanisms for redress when AI systems fail. It is also vital to delineate the roles and responsibilities of AI developers and law enforcement officers to prevent any abdication of responsibility.

Moreover, there are significant privacy concerns associated with the use of AI in surveillance. Advanced AI systems capable of facial recognition and predictive policing can infringe on individual privacy rights. The mass collection and analysis of data might lead to a surveillance state where the freedom to live without unwarranted monitoring is compromised. Ethical use of AI in surveillance necessitates stringent data protection measures, strict adherence to privacy laws, and ensuring that surveillance practices are proportionate to the intended security benefits.

The principle of informed consent is also relevant in the context of AI in law enforcement. Individuals should be aware of and consent to the use of AI technologies that affect their lives. In many instances, people are unaware that AI systems are being used to make decisions about them. Ensuring informed consent involves educating the public about the capabilities and limitations of AI and providing avenues for individuals to opt-out or challenge AI-driven decisions.

Furthermore, ethical considerations must address the potential for AI technologies to be misused by corrupt or malicious actors within law enforcement. The immense power of AI requires checks and balances to prevent abuses of power. Robust oversight mechanisms, regular audits, and clear anti-corruption policies are essential to safeguard against the misuse of AI. Establishing an independent oversight body to monitor the deployment and use of AI in law enforcement can provide an additional layer of accountability and protection against abuse.

Another critical aspect is the potential for AI to create dependency within law enforcement agencies. Relying too heavily on AI systems might erode the critical thinking and decision-making skills of human officers. While AI can serve as an invaluable tool, it should complement rather than replace human judgment. Continuous training and education programs for law enforcement officers are necessary to ensure that they remain adept at making informed decisions, especially in complex and nuanced scenarios where AI might fall short.

Finally, the ethical deployment of AI in law enforcement must consider the long-term societal impacts. The use of AI has the potential to reshape the social contract between citizens and the state. As technology evolves, it is essential to engage in ongoing ethical reflection and dialogue to understand the broader ramifications of AI on justice, equality, and human rights. Policymakers, technologists, ethicists, and community representatives must work collaboratively to develop and enforce ethical guidelines that keep pace with technological advancements.

Implementing ethical AI in law enforcement is not a one-time effort but a continuous process requiring vigilance and adaptability. Ethical principles should be interwoven into every stage of AI system development, deployment, and monitoring. By addressing these ethical implications head-on, we can harness the potential of AI to create a safer, fairer, and more just society.

Guidelines for Responsible Use

One of the most significant and controversial applications of AI today is its use in law enforcement. Navigating this complex landscape requires clear and rigorous guidelines to ensure that AI technologies are used responsibly. The purpose of these guidelines is threefold: to mitigate ethical concerns, protect individual privacy, and ensure that the deployment of AI systems aligns with societal values.

Firstly, transparency is non-negotiable when it comes to deploying AI in law enforcement. Law enforcement agencies must be forthright about where and how they use AI systems. This level of transparency builds trust within communities and helps to diminish fears of an Orwellian surveillance state. Clear communication regarding the types of data collected, the purpose for which these data are used, and the rights of individuals can alleviate public concern and foster a cooperative environment.

Moreover, developers and users of AI in law enforcement need to be vigilant about data quality and integrity. The lifeblood of any AI system is its data. If the data are biased, incomplete, or otherwise problematic, the resulting actions taken by law enforcement could be unjust and harmful. To ensure responsible use, agencies must commit to using datasets that are expansive, diverse, and representative. This reduces the risk of biased outcomes that otherwise could disproportionately affect marginalized communities.

Additionally, rigorous audit mechanisms must be established. These audits will serve as checks and balances, ensuring that the AI systems continue to operate fairly and accurately. External audits, carried out by third-party experts, can be especially valuable. They provide an unbiased assessment of the system's performance and adherence to ethical guidelines. Furthermore, periodic audits should be mandated to keep up with evolving societal values and technological advancements.

Equally important is the principle of accountability. Law enforcement agencies must be answerable for the decisions made by their AI systems. This includes establishing a clear chain of responsibility, so if an AI system makes an error or displays a bias, there's a swift and transparent method for addressing the issue. Correspondingly, the legal framework surrounding AI in law enforcement should be robust and adaptive, guiding agencies on the ethical deployment and usage of these technologies.

In parallel, continuous training and education for law enforcement personnel are essential. As AI technology evolves, so too must the knowledge and understanding of the officers using it. Training programs should encompass not only the technical aspects of these systems but also the ethical and legal concerns associated with them. By regularly updating their skill sets and ethical understanding, law enforcement officers can execute their duties more effectively and responsibly.

Another pillar of responsible AI use in law enforcement is public consultation and community involvement. It's crucial for law enforcement agencies to engage with community stakeholders, civil rights groups, and the general public when planning the deployment of AI systems. This participatory approach can offer diverse perspectives, identify potential ethical blind spots, and garner public support. It also fosters a sense of shared ownership and responsibility regarding the impact of these technologies.

Next, robust safeguards must be enacted to protect individual privacy. The collection and use of data should be guided by strict data protection laws. Techniques like data anonymization, encryption, and minimization can significantly lessen privacy risks. Additionally, clear protocols should exist for data retention and disposal, ensuring that data are not held longer than necessary and are used solely for their intended purpose.

To supplement these efforts, real-time monitoring and adaptive controls are essential. These tools allow for the continuous assessment and modification of AI systems to address unforeseen challenges quickly. Adaptive controls, for instance, can be programmed to respond to specific triggers that might indicate ethical boundaries are being approached or crossed. These systems are dynamic, offering a flexible approach to maintaining ethical standards over the lifespan of the AI application.

Furthermore, public awareness campaigns are invaluable. The general public needs to be educated about the capabilities, limitations, and intended uses of AI in law enforcement. Often, misunderstandings and misinformation can fuel fear and resistance. Transparent and effective communication can dispel myths and build a more informed and supportive public.

Despite these rigorous protocols, it is also essential to understand the limitations of AI. Over-reliance on technology can result in human complacency. There must always be a human element in decision-making, to evaluate the context and apply judgment that AI systems are not yet capable of. This human oversight helps to catch errors that an AI system might miss and ensures that ethical considerations are consistently applied.

Finally, cross-jurisdictional cooperation and standards are vital. Since AI systems used in law enforcement often operate across various jurisdictions, harmonizing ethical standards and practices can prevent a patchwork of regulations that are hard to navigate. International cooperation can further contribute to developing a consensus-driven approach, ensuring that the application of AI in law enforcement upholds global human rights standards.

In conclusion, the responsible use of AI in law enforcement isn't just about mitigating risks; it's about maximizing benefits while staying true to ethical principles that prioritize human dignity and justice. By adhering to these guidelines, law enforcement agencies can leverage AI's remarkable potential while safeguarding the values and rights that underpin a democratic society.

CHAPTER 10:
GLOBAL IMPLICATIONS AND POLICY

As artificial intelligence continues to evolve, its global impact becomes increasingly profound, necessitating thoughtful policy strategies and international collaboration. Diverse nations approach AI ethics in ways that reflect their unique cultural, socio-economic, and political landscapes, making harmonized global standards both challenging and essential. Policies that prioritize transparency, fairness, and accountability can help mitigate the dangers associated with AI, such as bias and misuse, while maximizing its potential for global good. Additionally, fostering international dialogue and cooperation offers a pathway to shared understanding and unified action against the risks AI poses. Indeed, by combining the wisdom of global perspectives, we can work towards crafting a future where AI serves the collective interests of humanity, balancing innovation with the principles of justice and human dignity.

International Perspectives

As artificial intelligence (AI) technologies permeate every facet of our global society, the response from different regions illustrates a mosaic of ethical stances, regulatory approaches, and cultural nuances. Each nation's take on AI ethics is deeply embedded in its historical and socio-political context, creating a diverse landscape that reflects various priorities and values.

Consider Europe's approach. The European Union (EU) has positioned itself as a leader in ethical AI by implementing robust regulations. The General Data Protection Regulation (GDPR) is a pioneering example, emphasizing privacy and data protection. More recently, the European Commission proposed the Artificial Intelligence Act, focusing on risk management and accountability. The EU's framework prioritizes human dignity and aims to create trustworthy AI, setting a high bar for others to follow. This comprehensive approach has inspired similar legislative efforts worldwide, promoting values such as transparency, fairness, and accountability.

Across the Atlantic, the United States has taken a different tact. The U.S. approach is characterized by a combination of state-level initiatives and federal guidelines rather than overarching regulations. This decentralized method stems from the nation's emphasis on innovation and economic competitiveness. While there are frameworks like the AI Bill of Rights proposed by the White House, which addresses several ethical concerns, a nationwide binding regulation is still forthcoming. The U.S. ethos is heavily influenced by the tech industry's rapid pace, balancing ethical considerations with fostering technological advancements.

Meanwhile, in Asia, countries like China and Japan demonstrate varied approaches to AI ethics and policy. China's AI strategy is embedded within its broader socio-economic goals. The government's emphasis on AI development is clear, but this comes with stringent state oversight and regulation. China's New Generation Artificial Intelligence Development Plan outlines ethical guidelines centered on safety, privacy, and fairness, yet it must be viewed through the lens of state control and surveillance. This approach starkly contrasts with Western paradigms, highlighting different interpretations of ethics and personal privacy.

Japan offers a unique perspective, merging high-tech development with traditional values. The Japanese government's AI strategy is grounded in the concept of "Society 5.0," which envisions a super-smart society. Ethical considerations in Japan are influenced by a cultural backdrop that emphasizes harmony and collective well-being. This reflects in policies prioritizing social benefit and integration of AI into daily life while upholding rigorous ethical standards.

The African continent presents another dimension to the international AI ethics discourse. Africa's approach is gradually evolving, with efforts to balance innovation and ethical deployment gaining momentum. Countries like Kenya, South Africa, and Nigeria are increasingly active in this space. Initiatives such as Smart Africa are crucial, aiming to create a knowledge-based economy by harnessing AI's potential while addressing issues like data sovereignty and ethical applications. Furthermore, local AI ethics frameworks are emerging, tailored to their unique challenges and opportunities, offering valuable insights into creating inclusive and responsible AI systems.

Latin America also contributes to this diverse international perspective. Countries like Brazil and Argentina are actively exploring AI policies that address ethical considerations in a region marked by economic and social disparities. The emphasis tends to be on inclusive AI that can bridge gaps in education, healthcare, and other critical sectors. Latin American nations are advocating for AI that not only drives economic growth but also addresses systemic inequalities, ensuring that technological advancements benefit all societal segments.

Moving to the Middle East, we see countries like the United Arab Emirates (UAE) setting ambitious goals for AI deployment, framed by ethical considerations. The UAE's National AI Program aims to position it as a global leader by embedding ethical guidelines into its AI strategies. This includes considerations for data protection, privacy, and societal impact, reflecting the region's blend of rapid modernization and traditional values.

In Oceania, Australia and New Zealand are making significant strides in developing their ethical AI frameworks. Australia's AI Ethics Framework underscores principles such as fairness, accountability, and transparency. Likewise, New Zealand's approach focuses on trust and citizen participation, ensuring that AI technologies are developed inclusively and ethically. These nations highlight the importance of public trust and engagement in AI ethics discourse.

It's essential to note that international perspectives on AI ethics and policy are not static; they are dynamic and continually evolving. Countries influence each other through shared knowledge, collaborations, and adaptations of best practices. International organizations such as the United Nations, UNESCO, and the OECD play critical roles in fostering global dialogue and establishing common ground on AI's ethical and regulatory aspects.

This global exchange can lead to a harmonized approach, yet it must respect regional nuances and cultural contexts. The aim should be to create interoperable ethical standards that promote responsible AI worldwide. For instance, the OECD's AI Principles, endorsed by over 40 countries, provide a foundational framework while allowing for localized adaptations.

However, the challenges in achieving a cohesive international approach are significant. Disparities in technological capabilities, economic conditions, and socio-political environments mean that a one-size-fits-all solution is impractical. Instead, it necessitates a polycentric approach where ethical AI policies are both globally aligned and locally relevant.

Furthermore, the ethical implications of AI extend beyond national borders, necessitating a coordinated response to issues like cross-border data flows, international trade impacts, and global security threats. These transnational challenges require collective action, emphasizing the importance of international cooperation and strategic alliances.

A globally integrated yet regionally sensitive approach can help balance innovation with ethical stewardship. It allows for the development of AI technologies that respect human rights, promote social justice, and contribute to global well-being while acknowledging and incorporating diverse cultural perspectives.

In conclusion, the international perspectives on AI ethics and policy reveal a complex yet interconnected world striving to navigate the ethical challenges posed by rapidly advancing technologies. By fostering global dialogue and collaboration, we can build a future where AI serves humanity responsibly and equitably, respecting the rich tapestry of human diversity.

Policy Recommendations

As artificial intelligence continues to integrate more deeply into various aspects of society, the need for robust and dynamic policy frameworks has never been more critical. The global nature of AI development and deployment necessitates a collaborative approach to policy-making that takes into account the diverse cultural, economic, and legal landscapes of different nations. Below are several key policy recommendations aimed at addressing the ethical challenges and privacy concerns associated with AI. These recommendations consider the insights gained from the preceding chapters and focus on establishing guidelines that promote responsible and equitable AI development.

1. Establishing International AI Governance Bodies

One of the most pressing needs in the current AI landscape is the establishment of international bodies that can set global standards and guidelines for AI ethics and governance. These bodies should involve policymakers, technologists, ethicists, and community representatives from a wide range of countries. Such organizations would not only foster international cooperation but also ensure that global standards are regularly updated to reflect technological advancements. This harmonization is essential to prevent fragmented regulations that could stifle innovation or lead to ethical discrepancies across borders.

2. Promoting Transparency and Accountability

Transparency and accountability should be foundational principles in AI regulation. Governments must require AI developers to provide transparent disclosures about how their systems operate, including the data they rely on and the algorithms they use. Moreover, regulatory bodies should mandate periodic audits of AI systems to verify compliance with ethical guidelines and to identify any potential biases or risks. This level of scrutiny will help build public trust in AI technologies, which is crucial for their widespread acceptance and use.

3. Safeguarding Data Privacy and Ownership

Data serves as the backbone of AI systems, making it imperative to establish stringent data privacy laws that protect individual rights. Policies should ensure that individuals have control over their personal data, including the ability to understand, access, and request the deletion of their data from AI systems. Additionally, there should be clear guidelines on data ownership, ensuring that individuals retain ownership of their data even when it is used by AI developers. This approach not only protects privacy but also fosters a culture of transparency and accountability among AI stakeholders.

4. Addressing Bias and Ensuring Fairness

Policymakers must take proactive steps to mitigate bias in AI systems. This can be achieved by enforcing the use of diverse and representative datasets during the development phase and by requiring regular testing for biases. Additionally, regulations should mandate the inclusion of bias mitigation strategies in the AI design process. By addressing bias at multiple stages of AI development, policymakers can help ensure that AI systems are fair and do not perpetuate existing social inequalities.

5. Fostering Public Participation and Education

AI policies should not be formulated in isolation. Public participation should be encouraged through open forums, consultations, and public debates, allowing citizens to voice their concerns and contribute to policy-making. This participatory approach ensures that policies are reflective of societal values and are more likely to gain public support. Furthermore, governments should invest in educational programs aimed at increasing AI literacy among the general public, helping people understand both the potential benefits and risks associated with AI technologies.

6. Supporting Ethical AI Research and Development

Governments should allocate funding for research in ethical AI, supporting projects that focus on developing models and systems that align with ethical guidelines. Incentives can also be provided for private companies that prioritize ethical considerations in their AI development. By financially supporting ethical research and development, policymakers can foster innovations that both advance technology and uphold societal values.

7. Creating Legal Frameworks for AI Accountability

CLEAR LAWS AND REGULATIONS ARE crucial for holding AI developers and users accountable for their actions. Legal frameworks should outline the responsibilities of AI developers, operators, and end-users, specifying liability in cases where AI systems cause harm. These frameworks must be adaptable, allowing for updates and modifications as the technology evolves. Additionally, the legal system should be equipped to handle AI-related cases with specialized training for judges and lawyers on AI issues.

8. Enhancing International Collaboration

AI is a global phenomenon, and addressing its challenges requires international collaboration. Countries must work together to share best practices, harmonize regulations, and develop joint strategies for managing the ethical implications of AI. International treaties and agreements can play a significant role in ensuring that AI technologies are developed and used responsibly, benefiting humanity as a whole.

9. Encouraging Corporate Responsibility

Corporations play a pivotal role in the AI ecosystem, and their commitment to ethical practices is essential. Policymakers should encourage companies to adopt ethical AI guidelines, such as the Asilomar AI Principles or similar frameworks. Corporate responsibility can be further incentivized through tax benefits, public recognition, or regulatory advantages. By holding companies accountable, governments can ensure that the private sector contributes positively towards ethical AI development.

10. Developing Ethical Guidelines for AI in Specific Sectors

Different sectors, such as healthcare, law enforcement, and finance, face unique ethical challenges when it comes to AI. Sector-specific ethical guidelines should be developed in consultation with industry experts, stakeholders, and ethicists. These guidelines will provide tailored solutions to address sector-specific risks while promoting the responsible use of AI technologies.

In conclusion, these policy recommendations aim to provide a comprehensive framework for addressing the ethical implications and privacy concerns associated with AI. By fostering international cooperation, promoting transparency, supporting ethical research, and ensuring public participation, we can create a balanced and fair AI ecosystem that benefits society at large. The development of AI technologies holds great promise, but it is through thoughtful and proactive policymaking that we can ensure this promise is fulfilled responsibly. The journey toward ethical AI is a collective endeavor, and it requires the commitment and collaboration of all stakeholders involved.

Collaborating for Global Solutions

In our interconnected world, the development and deployment of artificial intelligence (AI) are inherently global matters. The implications of AI cross national borders, affecting economic structures, cultural norms, and even the sovereignty of nations. This necessitates a comprehensive approach that involves international collaboration and cooperation. Establishing a framework for such global partnerships is no small feat, as it requires harmonizing policies and addressing disparities between nations. The goal is to foster an environment where the benefits of AI can be equitably shared while mitigating its risks.

International cooperation on AI policy can take many forms, ranging from bilateral agreements to multilateral platforms involving multiple countries. The European Union's General Data Protection Regulation (GDPR), for instance, has set a global benchmark for data privacy and protection standards. However, achieving similar consensus on AI guidelines is more challenging due to the varying degrees of technological advancement and disparate ethical frameworks across countries. Despite these challenges, international partnerships have the potential to bring about standardized ethical norms for AI development and deployment. This could ensure that AI technologies are used responsibly and for the greater good.

Collaborative frameworks also allow for the pooling of resources and expertise. Countries with advanced AI capabilities can support those with emerging sectors, facilitating knowledge transfer and technological assistance. This symbiotic relationship can help bridge the technological divide, enabling more equitable development. Organizations such as the United Nations and the World Economic Forum can play a crucial role in facilitating these partnerships, providing forums for dialogue and collaboration. Their efforts have already initiated global conversations around AI ethics, data privacy, and the urgent need for comprehensive regulatory frameworks.

Moreover, working together on global AI solutions can help address the biases that often creep into AI systems. Different cultures offer unique perspectives on fairness, ethics, and accountability, contributing to a more holistic understanding of these issues. When diverse voices are included in the discussion, it becomes possible to create AI systems that are more balanced and less likely to perpetuate existing societal biases. This aspect of collaboration is critical because biases in AI can have far-reaching consequences, particularly when these systems are used in critical sectors such as healthcare, law enforcement, and employment.

However, international collaboration is not without its challenges. Trust is a fundamental requirement for any partnership to succeed, and in the realm of AI, trust is often scarce. Countries might be wary of sharing their technological advancements or proprietary algorithms, fearing that such actions could compromise their competitive edge. Additionally, the lack of a standardized global framework for AI ethics complicates matters. The establishment of universal ethical standards could be a significant step toward building that trust. Such standards could be enforced through international treaties, much like those governing other global concerns such as climate change and nuclear proliferation.

One promising area for international collaboration is the development of "AI for Good" initiatives. These projects aim to leverage artificial intelligence to address global challenges such as poverty, climate change, and public health crises. By focusing on the common goal of improving human well-being, countries can find common ground and build trust. Initiatives led by the United Nations, like the Sustainable Development Goals (SDGs), offer a framework within which AI can be used to make meaningful contributions. Successful collaboration in these areas can serve as proof of concept, demonstrating that international partnerships can indeed lead to beneficial outcomes.

Equally important is the harmonization of legal frameworks to govern AI. Policies that are effective in one country might not be applicable in another due to different legal landscapes and societal values. International bodies can work towards creating adaptable policy frameworks that respect national sovereignty while ensuring global standards are met. For instance, the European Union's AI Act, which aims to regulate AI through a risk-based approach, provides a model that could be adapted for global use. If countries can align their regulations, it will enable smoother cross-border trade and innovation, benefiting the global AI ecosystem.

Furthermore, collaboration should extend to the research community. Shared research platforms and open-access publications can facilitate the dissemination of knowledge, allowing researchers from around the world to contribute to and benefit from scientific advancements. International conferences and collaborative research projects can help build a global community of AI experts who are committed to ethical practices. This academic collaboration can play a vital role in addressing the ethical and practical challenges posed by AI.

In conclusion, collaborating for global solutions in the realm of AI is both a necessity and a complex challenge. The rewards of such collaboration—enhanced technological advancements, equitable development, and ethical AI systems—are tremendous. It requires the concerted efforts of nations, international organizations, researchers, and policymakers. Overcoming the hurdles of trust and differing regulations will be difficult, but the potential benefits make it an endeavor worth pursuing. The future of AI depends on our ability to come together, share knowledge, and create frameworks that ensure artificial intelligence serves the greater good globally.

CHAPTER 11:
AI AND HUMAN RIGHTS

In the rapidly transforming landscape of artificial intelligence, the intersection of AI and human rights presents both unprecedented opportunities and profound ethical challenges. As AI systems become more integrated into various aspects of our lives, from facial recognition to decision-making algorithms, the imperative to uphold human rights becomes crucial. Ensuring that these technologies do not perpetuate or exacerbate existing inequalities requires a comprehensive understanding of fundamental human rights principles. The development and deployment of AI must consider the dignity, freedom, and privacy of individuals, especially those from vulnerable populations. Moreover, AI can be a powerful tool in humanitarian efforts, capable of aiding in disaster response and supporting human rights monitoring. Striking a balance between innovation and ethical responsibility is paramount, as we envision a future where technology serves humanity's best interests while safeguarding its fundamental rights.

Defining Human Rights in the AI Era

As artificial intelligence continues its rapid evolution, it brings both unprecedented opportunities and considerable risks, particularly when it comes to the realm of human rights. This is not merely a matter of theoretical debate but a pressing issue requiring concrete definitions and actionable frameworks. How we navigate this landscape will have profound influences on society's structure and individual liberties.

116

Human rights, at their core, are principles designed to protect the intrinsic dignity and freedom of every individual. Traditionally, these rights are guaranteed by local and international laws, ensuring protection against abuses like discrimination, oppression, and violence. However, the introduction of AI complicates these sacred guarantees. The technology can either bolster human rights through advancements such as improved healthcare and personalized education or erode them through mechanisms like surveillance and biased decision-making.

To define human rights in the AI era, it's imperative to start with an understanding of what constitutes a fundamental human right in this new context. Historically, documents like the Universal Declaration of Human Rights and the International Covenant on Civil and Political Rights have provided a solid foundation. Yet, AI introduces novel scenarios not previously contemplated by these frameworks. For instance, the right to privacy must be re-examined through the lens of mass data collection and predictive analytics, making it clear that human rights need to be adaptable."

One pressing issue is the right to privacy, which now intersects intricately with data ethics. In an age where data is a valuable commodity, the concept of informed consent becomes murky. Do individuals truly understand what they're consenting to when they agree to terms and conditions sprawling with legal jargon? This question underscores the need for redefining consent in ways that are transparent and comprehensible to the average person.

The right to autonomy also deserves redefinition. AI systems often make decisions that can significantly impact lives, from job recommendations to legal judgments. While these systems promise efficiency and objectivity, their opaque nature and the biases embedded within their algorithms can undermine individual autonomy. It's essential, then, that we establish guidelines ensuring these AI systems enhance rather than compromise personal agency.

Freedom from discrimination is another critical facet. Biased algorithms can perpetuate and even exacerbate existing societal inequalities. This is particularly concerning in sectors like criminal justice and finance, where biased AI systems can lead to unjust treatment. Take, for example, AI-powered hiring platforms that may inadvertently favor certain demographics over others. Mitigating this requires rigorous testing, transparency, and constant refinement aimed at fairness and inclusivity.

A more nuanced area involves the right to work and the potential impact of AI-driven automation on employment. While AI can streamline operations and foster innovation, it also threatens traditional job structures. The displacement of labor by machines poses ethical dilemmas concerning economic inequality and social welfare. Policies geared towards workforce retraining and the equitable distribution of AI's benefits are crucial to uphold this right.

Moreover, the alteration of human rights in the AI era necessitates active public participation in their redefinition. Community engagement and diverse stakeholder involvement are vital for crafting policies that are both inclusive and representative. Ignoring the perspectives of those who are most likely to be affected by AI-driven changes would result in a skewed understanding and potentially exacerbate existing inequities. Policymakers must foster environments where these voices can be heard and considered in all discussions surrounding AI ethics.

The concept of "digital dignity" is emerging as a cornerstone of modern human rights discussions. Digital dignity refers to the respect for individuals' rights and well-being in digital spaces. This includes protection against cyberbullying, data breaches, and manipulative tactics often employed by algorithms to influence behavior. Ensuring digital dignity demands a holistic approach, integrating tech design best practices and robust regulatory measures.

Another vital consideration is how AI can support and enhance human rights, particularly in humanitarian efforts. AI can efficiently monitor and address human rights violations, for example, through satellite imagery analysis to observe the displacement of communities or the use of natural language processing to detect hate speech. These applications demonstrate AI's potential to serve as a powerful tool for human rights advocacy, provided it is used responsibly and ethically.

Striking a balance between innovation and human rights is no easy task; it requires ongoing vigilance, interdisciplinary collaboration, and a proactive stance. Legal frameworks must be agile, adapting to technological advancements while remaining steadfast in their commitment to protecting human dignity. This not only involves enacting new laws but also updating existing ones to reflect the realities of the digital age.

Academic institutions and research organizations play a crucial role in this redefinition process. Their contributions to developing ethical AI systems provide a blueprint for aligning technology with human rights principles. Cross-disciplinary research that encompasses law, ethics, computer science, and social sciences is essential for identifying potential risks and creating robust safeguards.

Tech companies must also be transparent about how their AI technologies are developed and deployed. They have a responsibility to conduct due diligence, ensuring their products don't infringe on human rights. This includes implementing rigorous testing protocols, actively seeking to eliminate biases, and being open about their methodologies and data sources.

Finally, international cooperation is indispensable. Human rights are universal, transcending borders and cultures. A global approach to

AI ethics, endorsed by multinational agreements, can help harmonize regulations and ensure that the benefits of AI are distributed equitably. Collaborative efforts between nations can accelerate the development of standards and best practices that safeguard human rights while fostering innovation.

In summarizing, defining human rights in the AI era is a dynamic, multi-faceted challenge that calls for a revisited look at established principles through a contemporary lens. It involves the recognition of new rights, like digital dignity, and the modernization of traditional ones, such as privacy and freedom from discrimination. The continuous collaboration between policymakers, technologists, and society at large will be pivotal in this endeavor.

We stand at a crossroads where our decisions regarding AI can either fortify or diminish the essence of human rights. The AI era demands an evolved understanding, one that promises ethical advancements and a steadfast commitment to preserving the inherent worth and liberties of every individual. Establishing this balance will pave the way for a future where technology complements human values rather than competing with them.

Protecting Vulnerable Populations

The rapid advancement of artificial intelligence (AI) brings significant opportunities but also grave responsibilities. One of the most pressing responsibilities is protecting vulnerable populations. These are groups that, due to various socio-economic, political, or health-related reasons, are more susceptible to exploitation, discrimination, and harm. Addressing these issues requires a nuanced understanding of not just the technology, but also the societal contexts in which it operates.

Historically, technology has often been a double-edged sword for marginalized communities. On one hand, advancements can offer new pathways for education, healthcare, and economic opportunities. On the other hand, these same technologies can exacerbate existing inequalities or create new forms of harm. AI, with its ability to analyze large datasets and automate decision-making processes, holds the same dichotomous potential. Developers, policymakers, researchers, and anyone engaged with technology must pay close attention to how AI can both help and hurt vulnerable populations.

Understanding vulnerability in the context of AI involves identifying groups who may be adversely affected by algorithmic decisions or the deployment of AI systems. Vulnerable populations can include racial and ethnic minorities, elderly individuals, people with disabilities, immigrants, children, and those experiencing economic hardships. Each of these groups faces unique challenges that can be amplified by AI technologies.

A critical area of concern is bias in AI algorithms. Biases can stem from the data used to train AI systems, which often reflect historical and societal prejudices. For instance, facial recognition systems have been documented to be less accurate for people of color, perpetuating racial discrimination. This not only affects the accuracy of these systems but also their fairness and reliability, leading to potential injustices.

Moreover, predictive policing algorithms, which are used to allocate law enforcement resources based on crime data, can reinforce discriminatory practices. If the data used to train these systems are biased, the resulting AI models are likely to perpetuate those biases. This can result in a disproportionate targeting of already marginalized communities, further entrenching systemic inequities. Recognizing and addressing these biases is paramount to ensuring that AI systems do not undermine the justice they are supposed to promote.

Vulnerable populations often experience a lack of access to critical services and resources. Here, AI has the potential to bridge some of these gaps by providing more efficient and tailored services. However, the deployment of AI in areas such as healthcare and social services must be done with caution. In healthcare, for example, AI models that predict patient outcomes or recommend treatments could inadvertently favor populations that are already well-represented in the training data, thereby neglecting the needs of minority groups. This could exacerbate health disparities rather than ameliorate them.

The issue of consent is another crucial aspect when it comes to AI and vulnerable populations. Often, these groups lack the power or resources to give informed consent for their data to be used in AI systems. This raises significant ethical concerns about autonomy and the right to privacy. Policies and regulations need to be in place to ensure that data is collected and used responsibly, with robust mechanisms for consent that are inclusive and understandable to all individuals, regardless of their socio-economic or educational background.

To protect vulnerable populations effectively, it is essential to involve these communities in the AI development process. This means not only consulting them but also actively including their perspectives in the design, deployment, and monitoring phases of AI systems. Participatory design practices can help ensure that AI technologies meet the real needs of these communities while minimizing potential harms. Additionally, ongoing monitoring and evaluation are critical to promptly identifying and addressing any unintended consequences.

Regulatory frameworks play a vital role in safeguarding vulnerable populations from the potential harms of AI. Legislation should be enacted to enforce transparency, accountability, and fairness in AI systems. Existing human rights laws, such as those protecting against discrimination, should be adapted to address the nuances of AI technologies. Furthermore, international cooperation is necessary to develop global standards and guidelines that ensure AI benefits are shared equitably, and risks are mitigated worldwide.

Educational initiatives are also key to empowering vulnerable populations in the AI era. Providing accessible education and training on AI literacy can help individuals understand how AI affects their lives and how they can protect their rights. This includes not only formal education but also community outreach programs that raise awareness about AI's potential risks and benefits. Empowered individuals are better equipped to advocate for themselves and their communities.

While the ethical implications of AI are complex and multifaceted, the fundamental goal remains clear: to develop and deploy AI systems in ways that enhance human rights, rather than undermine them. This is a call to action for developers, policymakers, researchers, and all stakeholders to commit to ethical principles and practices that prioritize the protection of vulnerable populations. The advancement of AI should be a collective endeavor that seeks to uplift all of humanity, particularly those who are most at risk.

Ultimately, the protection of vulnerable populations requires a concerted effort from all sectors of society. It involves rethinking the ways we design, implement, and regulate AI technologies. By integrating ethical considerations at every stage of the AI lifecycle, we can create systems that not only avoid harm but actively contribute to the well-being and dignity of all individuals. This is not just a technical challenge; it is a moral imperative that calls on our shared humanity and collective responsibility.

The path forward involves embracing a holistic approach that combines technological innovation with robust ethical frameworks. It requires us to be vigilant, proactive, and committed to justice. By doing so, we can ensure that AI becomes a tool for empowerment, a force for good that transcends barriers and brings us closer to a world where everyone, regardless of their circumstances, can thrive.

AI and Humanitarian Efforts

The integration of artificial intelligence into humanitarian efforts marks a transformative era, reshaping the landscape of aid and emergency response. AI-driven technologies are being harnessed to address some of the most pressing global challenges, from natural disasters to refugee crises. By leveraging algorithms, data, and machine learning, organizations can now anticipate crises, allocate resources more efficiently, and even prevent crises before they escalate. The potential of AI in these domains introduces a new dimension of capability, bringing hope and unprecedented efficiency to humanitarian missions.

One of the primary areas where AI is making a significant impact is in disaster response. Predictive analytics can forecast natural calamities such as tsunamis, earthquakes, and hurricanes with increasing accuracy. These predictions enable organizations and governments to prepare and respond more effectively, potentially saving countless lives. For instance, machine learning models can analyze satellite imagery and weather data to predict the path of storms and identify vulnerable areas that require immediate attention. This proactive approach allows for the timely evacuation of communities and the strategic pre-positioning of resources, ensuring that aid reaches affected areas swiftly.

In addition to disaster prediction, AI is also enhancing the efficiency of relief operations. When a crisis occurs, the immediate priority is to assess the damage and identify the most urgent needs. AI-powered drones and computer vision technology can survey affected areas rapidly, providing real-time data to decision-makers. These tools can identify damaged infrastructure, locate survivors, and even detect potential health hazards such as contaminated water or blocked roads. This information is crucial for coordinating effective relief efforts and ensuring that aid is distributed where it is needed most.

Another critical application of AI in humanitarian efforts is in the realm of food security. Machine learning algorithms can analyze a vast array of agricultural data to predict crop yields and identify regions at risk of food shortages. By early identification of potential crises, humanitarian organizations can take preemptive action to mitigate the impact, such as by distributing food supplies in advance or implementing programs to support local farmers. Furthermore, AI can assist in optimizing the distribution of food aid, ensuring that it reaches the most vulnerable populations without delays or inefficiencies.

Refugee management is another area where AI demonstrates significant promise. With millions of people displaced by conflict and persecution, managing refugee populations pose immense logistical challenges. AI can streamline the registration process, track the movement of refugees, and allocate resources efficiently. For example, biometric technologies such as facial recognition can be employed at refugee camps to quickly and accurately register individuals, ensuring that they receive the necessary support. Additionally, AI can assist in identifying and reuniting families separated during their displacement, providing a semblance of stability and security amidst the chaos.

Health crises, such as the recent COVID-19 pandemic, further underscore the importance of AI in humanitarian efforts. AI-driven models can predict the spread of infectious diseases, enabling public health officials to implement containment measures effectively. Moreover, AI can assist in the rapid development of vaccines and treatments by analyzing vast datasets of scientific research and clinical trials. During outbreaks, AI can also optimize the distribution of medical supplies and personnel, ensuring that healthcare facilities are adequately equipped to handle surges in patient numbers.

Despite the vast potential of AI in humanitarian efforts, it is essential to address the ethical and practical challenges that accompany its deployment. One of the primary concerns is ensuring that AI-driven interventions do not exacerbate existing inequalities or biases. For instance, predictive models must be designed to consider the needs of marginalized communities, who are often disproportionately affected by crises. Furthermore, transparency and accountability in AI decision-making are crucial to maintaining trust among affected populations and humanitarian organizations. Stakeholders must have clear guidelines and oversight mechanisms to ensure that AI technologies are used responsibly and ethically.

Another challenge lies in data privacy and security. Humanitarian operations often involve the collection and processing of sensitive information about individuals and communities. It is imperative to safeguard this data to prevent misuse or unauthorized access. Robust data protection measures and compliance with international data privacy standards are essential to maintaining the integrity of humanitarian efforts and protecting the rights of those involved.

Collaborative efforts and partnerships are vital for maximizing the impact of AI in humanitarian contexts. Governments, non-governmental organizations (NGOs), tech companies, and academic institutions must work together to develop and implement AI-driven solutions. Sharing knowledge, resources, and expertise can accelerate innovation and ensure that technological advancements benefit the most vulnerable. Collaborative platforms and open-source projects can also foster a culture of transparency and accountability, promoting ethical AI practices across the board.

Education and capacity-building are equally important in leveraging AI for humanitarian efforts. Training programs and workshops can equip humanitarian workers with the skills needed to effectively use AI tools and interpret data-driven insights. Additionally, fostering a culture of continuous learning and adaptation can help organizations stay abreast of technological advancements and evolving ethical standards. Providing opportunities for communities to engage with and understand AI technologies can also enhance trust and cooperation, making humanitarian interventions more effective and inclusive.

Finally, it is imperative to highlight the stories of success and innovation in AI-driven humanitarian efforts. Case studies and real-world examples can serve as powerful motivators, inspiring others to explore the potential of AI in addressing global challenges. Sharing these narratives can also highlight the importance of ethical considerations and responsible practices, demonstrating that technological advancements and humanitarian principles can go hand in hand.

In conclusion, AI offers transformative potential for humanitarian efforts, providing tools that can predict crises, optimize resource allocation, and enhance the efficiency of relief operations. However, the deployment of AI in these contexts must be guided by ethical considerations, transparency, and collaboration. By addressing the challenges and maximizing the opportunities, we can harness the power of AI to create a more resilient and equitable world, where humanitarian efforts are more effective and inclusive than ever before.

CHAPTER 12:
DESIGNING ETHICAL AI SYSTEMS

As society moves towards an era deeply intertwined with artificial intelligence, the need to design ethical AI systems has never been more urgent. It's not simply about avoiding pitfalls; it's about creating frameworks that integrate ethical principles from the ground up. This involves aligning AI development with values like fairness, accountability, and transparency to ensure technology serves humanity responsibly. Developers and policymakers must collaborate to operationalize ethical guidelines, transforming abstract principles into actionable criteria within AI algorithms. Through real-world applications, ethical AI can demonstrate its potential to uplift societies while minimizing harm. By fostering an environment of continuous ethical vigilance and adaptation, we can build AI systems that are not only technically proficient but also morally sound, ensuring that innovation enhances the human experience rather than detracting from it.

Principles of Ethical Design

Designing ethical AI systems isn't just a technical challenge. It's a moral imperative. As artificial intelligence increasingly permeates every aspect of our lives, the principles guiding its design must be rooted in a commitment to human well-being, fairness, and justice. Making AI systems ethical involves adopting principles that ensure these technologies serve humanity, and not the other way around. Ethical design is, in essence, about embedding values into the very fabric of AI systems, making sure these systems act responsibly and equitably.

One of the foundational principles of ethical design is **transparency**. In a world where AI systems make decisions that can significantly impact individuals and society, openness becomes a necessity. Transparency involves more than just providing information about how AI works. Developers must ensure that AI systems are understandable. This means crafting models that are explainable and whose decision-making processes can be scrutinized. AI systems shouldn't be black boxes but rather should offer clarity about how inputs are transformed into outputs.

This leads us to **accountability**. AI systems must be designed with mechanisms that allow for accountability at every level. If an AI system makes a mistake, there needs to be a clear framework for identifying what went wrong and who is responsible. This principle ensures that AI developers and organizations can't shirk responsibility by hiding behind the complexity of their systems. Accountability also extends to the deployment phase, where constant monitoring and evaluation are necessary to ensure that AI systems remain aligned with ethical standards.

Fairness is another critical principle. AI systems can inadvertently perpetuate and even exacerbate biases present in the data they're trained on. It's incumbent upon designers to actively seek out and mitigate such biases. This means not only testing AI systems extensively but also adopting diverse datasets that reflect the multifaceted nature of human experiences. Fairness also involves ongoing assessments and updates to the AI systems, ensuring they evolve to be more just and equitable over time.

Prioritizing **privacy** is non-negotiable in ethical AI design. AI systems often require vast amounts of data to function effectively, but this data must be handled with the utmost care. Ethical design principles dictate that user data should be collected transparently, with explicit consent, and should only be used for the intended purposes.

Moreover, robust data protection mechanisms should be in place to prevent unauthorized access and ensure privacy throughout the data lifecycle.

Security is another key principle. Ethical AI systems must be designed to withstand various threats and attacks. This involves implementing robust security measures to protect both the AI systems and the data they process. It also means being proactive about identifying potential vulnerabilities and addressing them swiftly. The security principle is intertwined with the notion of robustness, ensuring that AI systems can perform reliably under a variety of conditions.

User autonomy should never be compromised. Ethical AI design ensures that users are given the ability to make informed decisions about how they interact with AI systems. This includes providing users with options to opt-in or opt-out and giving them control over their data. Empowering users with autonomy respects their agency and fosters trust in AI technologies.

Another crucial principle is **non-maleficence**, which is the commitment to "do no harm." This ancient ethical principle must be reincarnated in the digital age. AI systems must be built to minimize harm and should undergo rigorous testing to ensure they don't produce unintended negative consequences. This principle extends to considering the societal impacts of AI and striving to design systems that contribute positively to the community.

Beneficence goes hand-in-hand with non-maleficence. While the latter focuses on avoiding harm, beneficence emphasizes actively doing good. Ethical AI design should aim to enhance human well-being, improving quality of life and promoting positive outcomes. This could mean using AI to solve pressing global challenges or to provide support in areas like healthcare and education. The goal is to harness the power of AI for societal benefits.

Sustainability is increasingly relevant in today's world. Ethical AI design isn't just about immediate impacts but also considers long-term effects. AI systems should be designed in a way that ensures they are sustainable both in terms of their ecological footprint and their societal impacts. This means adopting energy-efficient algorithms and considering the broader environmental and social costs of AI deployment.

Inclusivity should underpin the entire design process. AI systems should serve and be accessible to all sections of society, regardless of their socioeconomic status, geographical location, or any other factor. This principle advocates for the creation of inclusive technologies that bridge gaps rather than widen them. Inclusivity also involves involving a diverse group of stakeholders in the design process to ensure a broad representation of perspectives.

Finally, **respect for human rights** is a non-negotiable foundation of ethical AI design. AI systems must uphold the dignity and rights of individuals, adhering to international human rights standards. This includes safeguarding freedoms such as privacy, expression, and association, and ensuring that AI technologies are not used to oppress or discriminate against anyone.

To sum up, the principles of ethical design are not merely abstract ideals. They are practical guidelines that inform every stage of AI development, from conceptualization to deployment. By adhering to these principles, we can create AI systems that not only advance technological capabilities but also uphold and promote ethical values. The design of ethical AI systems represents a partnership between humans and technology wherein both parties thrive and mutually benefit.

Implementing Ethical Guidelines

Designing ethical AI systems isn't just about having a clear vision or fancy algorithms; it's about embedding ethical guidelines into every stage of the development process. This approach ensures that AI systems not only function efficiently but also align with fundamental human values and societal norms. So, how do we achieve this? By weaving ethical guidelines into the very fabric of AI design, from ideation to deployment.

The first step in implementing ethical guidelines is to establish a robust ethical framework. This framework serves as a foundation, providing a set of principles and standards that guide decision-making throughout the AI development lifecycle. Important ethical principles might include fairness, accountability, transparency, and respect for privacy. Having a clear framework helps prevent potential ethical issues and enhances trust among users and stakeholders.

Once the ethical framework is established, the next step is to integrate these principles into the design and development processes. This involves developing clear, actionable guidelines that can be followed by engineers, designers, and researchers. Ethical considerations should be embedded in project planning, requirement gathering, and system design phases. These guidelines should be made accessible and understandable, so the entire team knows what's expected of them.

One of the most critical aspects of implementing ethical guidelines is to establish a diverse and inclusive team. Diverse perspectives can help identify potential ethical issues that a homogenous group might overlook. Encouraging diversity and inclusivity within the team may involve recruiting individuals from different backgrounds and expertise, including ethicists, sociologists, and domain experts. This multidisciplinary approach ensures that the AI system reflects a broad array of societal values and norms.

During the development phase, regular ethical reviews should be conducted. These reviews act as checkpoints, ensuring that the project is adhering to the established ethical guidelines. Ethical review boards, consisting of both internal and external experts, can provide valuable insights and feedback. This continuous assessment helps in mitigating risks and addressing ethical issues proactively rather than reactively, fostering a culture of ethical mindfulness.

Additionally, implementing ethical guidelines requires transparency and accountability in decision-making processes. Keeping detailed records of decisions, the reasoning behind them, and any ethical dilemmas encountered is vital. These records can serve as a reference for future projects and also contribute to the ongoing discourse on AI ethics. Transparency in decision-making not only strengthens internal validity but also builds external trust.

Another pivotal component is ethics training for all team members involved in the AI development lifecycle. Regular training sessions on ethical principles, real-world case studies, and the impact of AI on society can enhance awareness and vigilance. Training shouldn't be a one-time event but a continuous learning process, incorporating the latest ethical guidelines and societal expectations as they evolve.

Building user feedback mechanisms into the AI system can also play a significant role in maintaining ethical standards. User experiences and concerns provide valuable data on the system's ethical performance in real-world scenarios. Collecting, analyzing, and acting upon this feedback ensures that the AI remains aligned with user values and expectations, thereby fostering trust and acceptability.

Implementing ethical guidelines is not just a technical challenge but a cultural one. Creating an organizational culture that prioritizes ethics over expediency is paramount. This cultural shift begins at the top, with leadership setting the tone and establishing a clear mandate for ethical AI. Incentivizing ethical practices, rewarding employees and teams who uphold ethical standards, and openly discussing ethical dilemmas can fortify this culture.

Advanced technologies like AI often come with unpredictable risks, and a comprehensive risk management plan should be part of the ethical implementation. This plan identifies potential risks, assesses their impact, and outlines mitigation strategies. It also includes contingency plans for unforeseen ethical issues, ensuring the AI system can adapt and evolve while remaining ethically sound.

Furthermore, collaborating with external stakeholders, including policymakers, industry groups, and civil society, helps ensure that the AI system aligns with broader societal values and regulations. These collaborations can provide additional perspectives and help keep the AI system compliant with contemporary ethical standards and regulatory requirements. Open dialogues with external stakeholders promote transparency and trust, essential components of an ethical AI system.

Incorporating ethical guidelines into AI systems also means being vigilant about ongoing developments and ethical standards. The field of AI is rapidly evolving, and staying updated on the latest ethical frameworks, guidelines, and societal expectations is crucial. Continuous learning, attending workshops, and engaging with the broader AI ethics community can keep the team well-informed and ethically astute.

The ultimate goal of implementing ethical guidelines in AI design is to create systems that not only perform efficiently but do so in a manner that is responsible, fair, and beneficial to all stakeholders. By embedding these guidelines into every stage of the AI development lifecycle, we stand a better chance of achieving this goal, ensuring that AI technologies enhance human well-being and contribute positively to society.

Real-World Applications

The true test of designing ethical AI systems lies in their real-world applications. These applications span various domains—healthcare, law enforcement, finance, and beyond—each with its own set of ethical challenges. The aim is not just to conceptualize what an ethical

AI system looks like but also to see it functioning responsibly in diverse environments.

In healthcare, AI promises revolutionary advancements. AI algorithms analyze vast amounts of medical data faster and often more accurately than human doctors, enabling early disease detection and tailored treatment plans. But while these advancements offer hope, they also open doors to ethical dilemmas. For instance, the question of data privacy is paramount. Patients' medical histories are deeply personal, and misuse could have devastating consequences. Ensuring that AI systems comply with privacy laws like HIPAA in the United States becomes critical. Moreover, AI systems must avoid biases that could lead to inequalities in healthcare delivery. An AI trained predominantly on data from a particular demographic might fail to diagnose diseases accurately in underrepresented groups.

Law enforcement is another domain where the application of ethical AI is intensely scrutinized. AI-powered surveillance tools, predictive policing algorithms, and facial recognition systems can potentially reduce crime rates and improve security. However, these applications can also invade individual privacy and carry significant risks of racial and socio-economic biases. For example, facial recognition technologies have notoriously higher error rates when identifying people of color. To design ethical AI in this space, it's essential to ensure transparency, accountability, and rigorous testing against biases. Ethical guidelines should mandate regular audits and inclusive datasets that cover diverse populations.

Financial institutions are leveraging AI to monitor transactions for fraudulent activity, manage risks, and provide personalized financial advice. While these technologies enhance efficiency and security, they also bear ethical considerations. For example, AI systems might inadvertently reinforce existing biases in credit scoring or loan approvals, disadvantaging marginalized communities. Therefore, transparency in how these algorithms make decisions is crucial. Financial institutions must also establish mechanisms to allow individuals to challenge and correct discrepancies. Building robust AI systems that prioritize fairness can help reduce inequalities and foster trust in financial services.

The field of autonomous vehicles exemplifies another pressing real-world application. Self-driving cars rely on complex AI systems for navigation, obstacle detection, and decision-making in critical situations. The ethical design of these systems involves making life-and-death decisions, such as choosing to avert an accident by putting the car's occupants at risk or harming pedestrians. This raises substantial moral questions and requires a framework for ethical decision-making in split-second scenarios. Transparency in the algorithms' decision processes and accountability regarding who is held responsible in case of an accident are vital considerations.

In education, AI systems are utilized for personalized learning experiences, automating administrative tasks, and even guiding career counseling. Ethical design in educational AI must address concerns related to data privacy, consent, and the potential for perpetuating biases. For instance, an AI system that tracks students' performance and offers recommendations must ensure that the data handling complies with privacy laws. Moreover, these systems should be both transparent and explainable, so educators and students understand how decisions and recommendations are made. A balanced approach to data usage and ethical constraints can foster a more inclusive and fair educational environment.

AI in customer service has revolutionized the way companies interact with their customers. Virtual assistants and chatbots provide around-the-clock service, handling inquiries efficiently. However, designing these systems ethically involves ensuring they don't inadvertently cause harm or frustration. For instance, customers should be made aware that they are interacting with an AI system and have the option to speak with a human representative if desired. Additionally, these systems must be designed to handle sensitive information responsibly, safeguarding user data against unauthorized access.

Humanitarian efforts also benefit from ethical AI applications. AI systems can assist in disaster response, resource allocation, and predict potential crisis zones. For example, AI can analyze data from various sources to predict earthquakes, floods, or other natural disasters, allowing for preemptive action and efficient resource distribution. However, ethical considerations such as data sovereignty, privacy, and the potential for misuse in conflict zones must be weighed carefully. Collaborations with local communities and ensuring transparency in how data is collected and used can help mitigate these concerns.

The entertainment industry utilizes AI for content recommendation systems, automated content creation, and even deepfake technology, which has both creative and destructive potential. Ethical AI design here must address the question of intellectual property, consent, and the potential for misinformation. Content recommendation systems should avoid creating echo chambers that limit user exposure to a diverse range of ideas and perspectives. Similarly, while deepfake technology offers innovative storytelling possibilities, it also raises significant ethical concerns about authenticity and the potential for misuse in spreading false information. Establishing clear ethical guidelines and robust verification systems can help navigate these challenges.

In the realm of environmental conservation, AI systems play a pivotal role. They assist in monitoring wildlife populations, predicting environmental changes, and managing natural resources sustainably. Ethical design here ensures that data collection methods do not harm wildlife and that AI models are used to promote conservation efforts rather than exploit natural resources. Collaboration with environmental scientists and local communities can enrich the data and algorithms, making AI applications more effective and ethically sound.

Retail and e-commerce have seen a surge in AI applications for inventory management, personalized shopping experiences, and supply chain optimization. Ethical considerations in this sector include data security, transparency in personalized recommendations, and the potential impacts on employment. For instance, while AI can optimize inventory and reduce waste, there must be transparency about how consumer data is utilized. Additionally, companies should consider the socio-economic impact of AI-driven automation on retail workers and explore ways to mitigate job displacement.

In governance and public policy, AI systems can enhance decision-making processes, optimize public resource allocation, and improve citizen services. Ethical AI design in government applications must prioritize transparency, accountability, and inclusivity. Decisions made by AI should be explainable, ensuring that citizens understand how policies are formulated and implemented. Public consultations and stakeholder engagements can ensure that AI systems serve the broader community's interest and uphold democratic values.

The military sector's adoption of AI presents some of the most significant ethical challenges. From autonomous drones to AI-driven cybersecurity systems, the potential for misuse and the moral implications are immense. Ethical design in military AI must consider the rules of engagement, adherence to international laws, and the potential for unintended consequences. Clear ethical frameworks and international cooperation are necessary to ensure that AI technologies are used responsibly and do not escalate conflicts or cause unnecessary harm.

Ethical AI systems in journalism and media can help combat misinformation, enhance investigative journalism, and provide personalized content while respecting user privacy. However, the challenge lies in balancing the need for personalized media with the risk of creating information bubbles. AI algorithms should be designed to prioritize factual accuracy and diverse viewpoints, ensuring a well-informed public.

The agriculture industry benefits from AI through precision farming, crop monitoring, and climate prediction. Ethical considerations include protecting farmers' data privacy, ensuring equitable access to AI technologies, and promoting sustainable farming practices. Collaboration with agricultural experts and local farming communities can enhance the ethical deployment of AI in agriculture.

Ultimately, the real-world applications of ethical AI systems span a multitude of sectors, each with unique challenges

Conclusion

The culmination of our journey through the ethical dimensions of artificial intelligence brings us to a profound realization: AI represents both an unparalleled opportunity and an equally formidable challenge. As we stand at the precipice of a new era driven by intelligent systems, it is clear that the ethical framework we establish today will shape the future of human-AI interaction for generations to come.

Throughout this book, we've explored the multifaceted nature of AI ethics, diving into foundational concepts, privacy concerns, bias and fairness, transparency, accountability, and more. Each chapter has underscored the complexities and nuances involved in creating and managing AI systems that align with our ethical values. The diversity of perspectives provided—from technical insights to broader societal implications—reflects the rich tapestry of considerations that must be woven together to form a cohesive ethical strategy.

One of the primary takeaways is the critical importance of integrating ethical principles into the very fabric of AI development. This integration is not merely a box-ticking exercise but a commitment to fostering trust, fairness, and respect for human rights. The principles of ethical design, which include transparency, accountability, and fairness, must be embedded in AI systems from their inception, guiding every stage of the AI lifecycle.

Another crucial theme is the role of privacy and data protection. In an era where data is the lifeblood of AI, safeguarding individuals' privacy is paramount. We must not overlook the risks associated with data collection and use, as these can lead to significant breaches of trust and potential harm. Mitigating these risks involves implementing robust data protection measures and fostering a culture of respect for privacy.

Bias and fairness in AI systems have been thoroughly examined, highlighting the need for vigilance in detecting and addressing biases. As AI increasingly influences key aspects of our lives, from hiring practices to law enforcement and healthcare, ensuring that these systems operate equitably is non-negotiable. Practical strategies for reducing bias, alongside ongoing scrutiny and updates, are necessary to maintain the integrity of AI systems.

Transparency and explainability form another pillar of responsible AI. Users and stakeholders must understand how AI systems make decisions, especially in high-stakes scenarios. Explainable AI models contribute to this understanding, fostering trust and enabling accountability. Transparent communication about AI processes and decisions helps demystify the technology and promotes its acceptance and responsible use.

Accountability remains a cornerstone of ethical AI. As we've seen, defining clear accountability structures ensures that there is always a responsible entity in the event of misuse or failure of AI systems. Legal and ethical standards must evolve in tandem with technology to provide the necessary frameworks for accountability. Case studies have shown how accountability can be practically implemented, offering valuable lessons for future developments.

The impact of AI on the workforce, healthcare, security, and law enforcement has brought to light specific ethical dilemmas and potential benefits. Preparing for an AI-driven future in the workforce involves not only training and education but also rethinking employment paradigms to accommodate new roles and opportunities created by AI. In healthcare, while AI promises tremendous advancements, it also poses ethical dilemmas that must be carefully navigated to ensure patient safety and equity.

Security and robustness are essential to the resilience of AI systems. Threats to AI, both external and internal, can undermine their effectiveness and safety. Ensuring robustness involves designing systems that can withstand adversarial attacks and continue to function reliably. Security best practices are vital in safeguarding these systems and maintaining public confidence.

In the realm of law enforcement, the ethical implications of AI surveillance and predictive policing have been scrutinized. Clearly, there is a delicate balance between leveraging AI for safety and protecting individual rights. Responsible guidelines and regulations are crucial to prevent overreach and ensure that AI applications in law enforcement are fair and just.

On a global scale, AI policy and collaboration highlight the importance of international cooperation. Different cultural, legal, and ethical perspectives must be reconciled to create coherent global standards. Policy recommendations provided in this book aim to bridge gaps and foster collaborative solutions that respect diverse viewpoints while achieving common goals.

Human rights in the AI era encapsulate the overarching need to protect vulnerable populations and promote humanitarian efforts. AI must be leveraged to uphold and advance human rights, rather than infringing upon them. Designing ethical AI systems with these considerations at the core can lead to transformative positive impacts globally.

In conclusion, the path to ethical AI is a continuous journey requiring the collective efforts of developers, policymakers, researchers, and society at large. By committing to ethical principles, fostering transparent practices, and upholding accountability, we can harness the immense potential of AI while mitigating its risks. As we advance, the lessons and strategies discussed in this book will serve as a compass, guiding us towards a future where AI not only augments human capabilities but does so in a manner that is just, equitable, and respectful of our shared ethical values.

Our responsibility is clear: to cultivate an ethical AI landscape that not only benefits our generation but also paves the way for a more inclusive, fair, and transparent future. This is our moment to shape AI governance wisely and prudently, ensuring that our technological advancements remain steadfast allies in the quest for a better world.

APPENDIX A:
APPENDIX

The objective of this Appendix is to consolidate essential information and supplementary materials that support the main text of the book. Here, you will find additional insights and resources that are critical for a comprehensive understanding of the ethical challenges and privacy concerns in artificial intelligence. Each section within this appendix aims to enrich your grasp on the topics discussed in the preceding chapters, providing you with tools and references to foster responsible AI development.

Key Concepts and Definitions

A solid understanding of the fundamental concepts discussed throughout the book is crucial for navigating the complex ethical landscape of AI. Below is a summary of key terms and definitions that are often referenced.

Artificial Intelligence (AI): The capability of a machine to imitate intelligent human behavior, accomplishing tasks such as visual perception, speech recognition, decision-making, and language translation.

Bias: Inclinations or prejudices for or against something or someone, which can affect the fairness of AI systems.

Explainability: The extent to which the internal mechanics of a machine or deep learning system can be explained in human terms.

Robustness: The ability of an AI system to perform reliably under a variety of conditions.

Transparency: The openness and clarity with which AI systems operate, allowing stakeholders to understand how decisions are made.

Supplementary Readings and Resources

To further your knowledge and provide a broader context, consider the following readings and resources:

- *Books*: Works by authors who have significantly contributed to the field of AI ethics.

- *Research Papers*: Academic papers that delve into detailed studies and findings relevant to AI ethics and privacy concerns.

- *Online Courses*: MOOCs and other educational programs that offer structured learning on AI ethics and responsible development methodologies.

- *Government and NGO Reports*: Insights from public policy reports and whitepapers that address global challenges and strategic initiatives in AI ethics.

Ethical AI Guidelines and Standards

Several organizations and institutions have developed guidelines and standards to promote ethical AI development. These resources can serve as valuable references for implementing best practices:

- IEEE's "Ethically Aligned Design"
- EU's "Ethics Guidelines for Trustworthy AI"
- Partnership on AI's "Tenets"

Case Studies

Real-world case studies provide insight into the application of ethical principles in AI development and deployment. They illustrate both challenges and successful practices, offering practical examples:

- **Healthcare AI Implementation**: Examining how AI has been used to enhance diagnostic accuracy while addressing privacy and ethical concerns.

- **AI in Law Enforcement**: Analyzing the balance between efficacy in surveillance and the potential for bias and infringement of human rights.

- **Workplace Automation**: Investigating the impacts of AI-driven automation on employment, including strategies for workforce adaptation.

Practical Tools and Frameworks

The following tools and frameworks can assist developers and organizations in building ethically responsible AI systems:

- *Fairness Indicators*: Tools that help measure the fairness of machine learning models.

- *Explainability Toolkits*: Libraries and software that provide insights into how models make decisions.

- *Security Checklists*: Best practices for ensuring the robustness and security of AI systems.

This appendix, while comprehensive, is by no means exhaustive. It aims to be a starting point for deeper exploration and continual learning in the evolving field of AI ethics. Your commitment to understanding and addressing these issues is a vital step towards developing AI systems that are not only innovative but also equitable and just.

GLOSSARY OF TERMS

This glossary serves as a quick reference, offering precise definitions to some of the key terms commonly encountered in discussions on the ethical implications of artificial intelligence. Understanding these terms will help you navigate through the complexities of AI and its ethical landscape.

Algorithm

A set of rules or instructions designed to solve a problem or perform a specific task. In AI, algorithms are used to process data and make decisions.

Artificial Intelligence (AI)

The simulation of human intelligence processes by machines, particularly computer systems. These processes include learning (acquiring information and rules for using the information), reasoning (using the rules to reach conclusions), and self-correction.

Bias

A predisposition or preconceived notion that affects the neutrality of an AI system. Bias can arise from the data used to train AI models or from the algorithms themselves.

Black Box

A system or model whose internal workings are not visible or understandable to the user. In AI, black-box models make decisions that are difficult to interpret or explain.

Data Privacy

The aspect of information technology that deals with the ability to control how data is collected, shared, and used. In the context of AI, data privacy is crucial for protecting individuals' personal information.

Ethical AI

The practice of designing AI systems that are aligned with ethical principles and social values. This includes considerations of fairness, transparency, accountability, and human rights.

Explainability

The ability of an AI system to provide understandable and interpretable outputs or decisions. Explainability is vital for gaining trust and ensuring accountability in AI systems.

Machine Learning (ML)

A subset of AI that enables systems to learn from data and improve their performance over time without being explicitly programmed. ML algorithms build models based on sample data to make predictions or decisions.

Privacy Risks

The potential threats to individuals' personal information resulting from data collection, storage, and analysis. In AI, privacy risks can arise from the misuse of data or lack of proper security measures.

Robustness

The ability of an AI system to maintain its performance and reliability under various conditions, including unexpected inputs or adversarial attacks. Robust systems are crucial for ensuring safety and security.

Transparency

The attribute of being open and clear about how AI systems work, including their data sources, decision-making processes, and limitations. Transparency fosters trust and allows for better scrutiny and accountability.

Vulnerable Populations

Groups or individuals who are at greater risk of harm or exploitation due to their socio-economic, demographic, or health status. In AI ethics, protecting these populations is essential to ensuring equitable outcomes.

Additional Resources

For those who want to dive deeper into the ethical implications of artificial intelligence, a wealth of additional resources is available to augment your understanding. Whether you're a developer, policy maker, researcher, or tech enthusiast, having access to comprehensive materials can provide you with the nuanced perspectives needed to navigate this complex field.

A fundamental cornerstone of building ethical AI systems involves staying current with ongoing research and thought leadership. Journals such as the *Journal of Artificial Intelligence Ethics* and conferences like the AAAI/ACM Conference on AI, Ethics, and Society are excellent starting points. These resources offer peer-reviewed articles, case studies, and expert opinions that will broaden your understanding and keep you updated on the latest trends and emerging challenges in AI ethics.

Books are another invaluable resource. For a rigorous, academic exploration, titles such as *Ethics of Artificial Intelligence and Robotics* and *Artificial Intelligence: A Modern Approach* provide foundational knowledge. Additionally, more accessible reads like *Radical Technologies: The Design of Everyday Life* and *Weapons of Math Destruction* offer engaging discussions about the societal impacts of AI. Each of these books presents diverse perspectives, helping readers grasp the broad scope of ethical considerations in AI development.

Government and non-profit organizations also offer extensive resources. Regulatory frameworks and guidelines issued by entities like the European Commission's High-Level Expert Group on Artificial Intelligence, and reports from the Future of Life Institute, provide practical advice on creating responsible policies. These resources are invaluable for policy makers looking to draft regulations that balance innovation with ethical responsibility.

Online courses and webinars are particularly beneficial for those seeking more structured learning experiences. Platforms like Coursera, edX, and Udacity offer tailored courses on AI ethics, many of which are created by leading universities and industry experts. Topics covered range from introductory concepts to advanced methodologies, providing a comprehensive curriculum that suits all experience levels.

For developers, toolkits and frameworks designed to foster ethical AI practices are essential. Resources like IBM's AI Fairness 360 Open

Source Toolkit and Google's What-If Tool empower developers to detect and mitigate bias, ensure transparency, and promote accountability in AI systems. These resources often come with extensive documentation, tutorials, and community support, making them useful for both novice and experienced developers.

Academic institutions and research labs also contribute significantly to the AI ethics landscape. Universities such as MIT, Stanford, and Oxford have dedicated centers and labs focused on the intersection of AI and ethics. The research produced by these institutions often sets the stage for public discourse and policy decisions, making their studies and publications important resources for anyone involved in AI.

For those interested in the global implications and policy aspects of AI, international organizations such as the United Nations and the World Economic Forum publish influential reports and white papers. These documents often discuss the broader societal impacts of AI, offering a holistic view that goes beyond the technical details to consider geopolitical and cultural dimensions.

It's also beneficial to engage with community-based organizations and advocacy groups. Entities such as the Algorithmic Justice League and Data & Society conduct critical work at the grassroots level, pushing for greater transparency, fairness, and accountability in AI systems. These groups often offer workshops, seminars, and action guides that can help individuals and organizations ensure their AI practices are ethically sound.

Lastly, for real-world applications and case studies, industry reports from companies like Microsoft, Google, and IBM frequently feature valuable insights. These reports often cover practical implementations, lessons learned, and best practices, serving as a pragmatic guide for those looking to apply ethical principles to their own AI projects.

In summary, the field of AI ethics is vast and multi-faceted, requiring a continuous commitment to learning and adaptation. By leveraging the wealth of resources available across various mediums—academic journals, books, online courses, toolkits, institutional reports, and community organizations—individuals and organizations alike can stay informed and act conscientiously in the development and deployment of AI technologies.

endnotes

In crafting an understanding of the ethical landscape surrounding artificial intelligence, this book sought to articulate clear frameworks, elucidate complex considerations, and provoke thought about the future we're collectively building. The "Glossary of Terms" is intended as a reference to help navigate the often intricate and jargon-laden discussions of AI and ethics. These endnotes serve as a comprehensive wrap-up, situating key terms within the larger narrative, underscoring their relevance and encouraging continued exploration.

The terms gathered here aren't isolated; they are deeply intertwined with the ethical, social, and technical fabric of AI systems. For developers, these definitions should serve as a touchstone for ethical design and decision-making. Each term isn't just a static definition but a dynamic concept with implications that ripple through all aspects of AI development and application. From "bias" to "transparency," these words capture the essence of the ethical challenges faced today and will likely continue to face in the future.

One of the critical insights gleaned from understanding these terms is the interconnectedness of concepts like fairness, accountability, and privacy. These aren't standalone issues that can be tackled in isolation but are part of a broader ethical tapestry. Comprehending the full spectrum of these terms can reveal how closely they are interwoven, emphasizing the importance of holistic approaches to AI ethics.

For policymakers, the glossary provides a foundation from which to build robust regulations and guidelines that can help steer AI development towards beneficence and justice. The clarity of terms like "explainability" and "robustness" can play a crucial role in forming legislation that accurately addresses the multifaceted challenges of AI. A shared understanding of these terms also fosters better communication and collaboration across sectors, aiding in the formation of policies that are both realistic and effective.

Researchers will find that these terms are a stepping stone for deeper inquiry. While the definitions serve to establish a common language, the real value lies in the nuanced discussions and inquiries they prompt. Each term can root a new research question or direction, setting the stage for innovative studies that probe the ethical implications more deeply. Consider the term "autonomy"—it's not merely about the independence of AI systems but about the human autonomy that's impacted by these systems.

For tech enthusiasts, understanding these terms can enrich the dialogue around AI and its societal impacts. Whether participating in discussions, contributing to open-source projects, or simply engaging with AI tools, having a grasp on terms like "surveillance" or "human rights" empowers more informed and impactful participation. This glossary is not just a list of definitions but a gateway to a richer, more nuanced understanding of AI ethics.

One of the key takeaways from the endnotes is the importance of continuous learning and adaptation. AI is a rapidly evolving field, and the associated ethical considerations will likewise evolve. The terms listed in the glossary might take on new meanings or reveal new dimensions as technology progresses. Thus, staying updated with the latest developments and revisions in terminology is crucial for anyone engaged in the field of AI ethics.

Moreover, the cross-references included in these endnotes highlight the multi-dimensional nature of ethical discussions in AI. A term like "accountability" isn't confined to a single chapter; it spans discussions of law, fairness, and design principles. Recognizing and embracing these intersections ensures a more comprehensive and integrated approach to addressing ethical issues.

It's also worth mentioning the deliberate inclusion of multiple perspectives in the glossary. Understanding how different cultures, political systems, and philosophical traditions interpret terms can lead to more inclusive and globally relevant AI systems. For instance, "privacy" might hold different connotations and legal interpretations in various parts of the world, and acknowledging these differences enriches the overall discourse.

Through these endnotes, we hope to reinforce the narrative that ethical AI is a collaborative effort. It requires the concerted efforts of developers, policymakers, researchers, and tech enthusiasts, all working towards common goals while appreciating their diverse roles. By grounding the discussion in well-defined, transparent terms, we aim to support collective action that is informed, intentional, and ethically sound.

Ultimately, the glossary and these endnotes are tools—they are not an endpoint but a beginning. They are designed to be a starting point for deeper exploration, critical thinking, and ongoing dialogue about the ethical implications of AI. The journey towards responsible AI isn't a solitary path but a shared voyage that demands vigilance, empathy, and mutual respect. With this foundation, we can move towards a future where AI serves humanity in ways that are just, equitable, and profoundly positive.

Looking ahead, the continuous refinement and expansion of the glossary will be essential. As new terms and concepts emerge, keeping this resource updated will empower all stakeholders to stay informed and engaged. It's a living document that reflects the dynamic nature of AI and its ethical landscape, encouraging perpetual learning and adaptation.

We encourage readers to return to these endnotes and the glossary often, using them as a reference point and guidepost. The terms within encapsulate key ethical principles and serve as milestones on the roadmap of responsible AI development. By thoroughly understanding these terms, stakeholders at all levels can make more informed decisions, foster meaningful dialogue, and contribute to a more ethical AI ecosystem.

In summary, the "Glossary of Terms" and these endnotes constitute essential resources for anyone invested in the ethical implications of AI. They are meticulously crafted to serve as guiding lights, illuminating the path towards responsible and humane AI. Let these tools support your ongoing efforts and inspire you to contribute to a future where technology and ethics coexist harmoniously.